British Field Crops

A Pocket Guide to the Identification, History and Uses of Traditional and Novel Arable Crops In Great Britain

Written and illustrated by
Sally A. Francis

M.A. (Oxon.) D.Phil. (Oxon.)

To Amy, Mark, Jo, Mum, Dad, Shirley & Ken.

Published in 2005 by

Sally Francis, The Coldens, Ingham Road, West Stow, Bury St Edmunds, IP28 6EX.

ISBN 0-9550466-1-0

Acknowledgements

This book would have been impossible to write without the generous help of the following who supplied information, seed samples, whole or parts of plants and/or allowed me to visit their crops: John Barrett, Dr Mike Field & Darryl Playford (Advanta Seeds UK), Steve Croxton (Biomass Industrial Crops Ltd.), Kevin Sawford (Broom's Barn Research Station), Tim Church & Geoff Lakin (W. A. Church (Bures) Ltd.), Robert Walpole (Colman's of Norwich), John Close & Lesly Lawton (DEFRA Stats Dept), Mike Craner (Harlow Agricultural Merchants Ltd.), John Hobson & Dan Squier (Hemcore Ltd.), Dr Ray Mathias (John Innes Centre) Julie Goult & Neal Boughton (John K. King & Sons Ltd.), Paul Nelson (National Institute of Agricultural Botany), Richard Alander (Nickerson Seeds Ltd.), David McNaughton (Soya UK Ltd.), Simon Meakin (Springdale Crop Synergies Ltd.), Alan Roffey (SW Seed Ltd.), Steve Corbett (Syngenta Ltd.), Philip and Lisa Haylock (J. & P. Haylock Farms, Beck Row), Paul Williamson (Church Farm, Bradfield Combust), Dominic Watts (Rushbrooke Farms, Rushbrooke), Paul Hurn (James Sutton Farming, Thorney), Gavin Davies (Chivers Farms Ltd., Impington), and the following individuals: Jo, Jill & Ted Francis; Dr Crawford Kingsnorth; Celia Jeal; and Will Foss.

Thanks go to the Action Desks at BBC Radio Norfolk and BBC Radio Suffolk that broadcast my requests to contact a grower of durum wheat, and to the following for their comments on the manuscript: Dr Philip Draycott, Tracey Welham, Jill Francis, Jo Garrard and Dr Mark Winterbottom.

Finally I would like to thank my family and friends for their continuing help and support while this book was being researched and written.

Printed by Moreton Hall Press 01284 767442

British Field Crops

Contents

How to use this book

This book is a guide to arable crops in Great Britain. With the exception of asparagus and miscanthus, each crop included here is grown as an annual, meaning that it occupies the land for a single growing season, at the end of which the land is ploughed.

This book is arranged into two directories, reflecting the two classes of the flowering plants: **grasses & related species** (monocotyledons; plants that produce one seedling leaf on germination), and **broadleaved species** (dicotyledons; plants that produce two seedling leaves on germination). A summary of characteristics that can be used to distinguish adult plants is in the following table.

Characteristic	Grasses etc.	Broadleaved plants
Leaf veins	Parallel	Net-like or branching
Roots	Usually fibrous	Some have a taproot
Flower structures, e.g. petals, stamens etc.	Usually in groups of three, or multiples of three	Often in threes, fives, or multiples of these

Within each class, plants are grouped into botanical families. In this book, their order reflects the relative area planted in Great Britain. Within each family, the species are also broadly arranged according to areas planted, but with similar-looking species together to make identification easier.

On each page of the crops directories, you will find illustrations and information about the crops. In the header of each page is the **botanical family** to which the crop belongs. There are also estimates of the **total area** grown in Great Britain.

■☐☐☐☐	Less than 5,000 ha
■■☐☐☐	5,001 to 20,000 ha
■■■☐☐	20,001 to 140,000 ha
■■■■☐	140,001 to 1,000,000 ha
■■■■■	Over 1,000,000 ha

Each **scientific name** is presented in full and includes, in non-italics, the name or initials of the 'author' of each name. Names are taken from Mabberley, D. J. (1997) *The Plant-Book*. CUP. Synonyms are also given, to allow the information in this book to be compared with other sources.

There is a **description** of the basic physical characteristics of the crop, whether it occurs in distinct forms, and the identifying features that can be used to distinguish it from similar plants. This is accompanied by botanical line drawings to illustrate the appearance and growth habit of each crop. Although there is a quick identification key based partly on flower colour on the following page, flower colour is not the primary characteristic used in this book, for two reasons: (i) several crops do not usually flower during a normal growing season, e.g. turnips and other root crops and (ii) a number of crops can have more than one different flower colour, e.g. potatoes can have mauve or white flowers, and the flowers of echium can be white or shades of blue and purple.

The **origin & history** of each crop reveals that most crops grown in Britain are not native species, but have been introduced from all over the world. Some species have been domesticated for thousands of years, during which they have diverged into distinctly different crops, e.g. the brassica vegetables such as cabbage, cauliflower and Brussels sprouts. Other crops are novel, and have only been commercialised during the last few years.

The **uses** of the crop and any by-products are listed.

The approximate dates on which crops are **sown** and **harvested** are features that can also help identify an unknown crop and give a rough idea of its growth stage, depending on the time of year.

Crops are **grown** in different geographical regions of Britain. Some are cultivated countrywide, whilst others are restricted by climate, soil requirements or proximity to specialist processing factories. This is another clue to help identify an unknown crop.

A note on terminology
Whilst every effort has been made to avoid using technical words, inevitably in some instances there is no alternative. These are defined in the Glossary. In the case of root crops, where the 'root' may comprise true root tissue and swollen stem or other tissue, the term 'root' is retained. Although in strict botanical terms the seeds of a plant are contained within a fruit, in this book the terms 'pod' and 'capsule' are used to avoid confusion with the every-day definition of 'fruit'. Additionally, to avoid confusion 'variety' has been used in place of the botanical term, 'cultivar'.

Crop identification key

1. Major cereals, based on spikelet structures (see p. 7 for drawing)

If there is/are...	and these other features...	it could be...
1 grain per spikelet	ears with long straight awns (bristles)	2-rowed barley, p. 9
1	drooping ears with a very large number of seeds	millet, p. 14
2	ears with short awns, tips of grains visible	rye, p. 10
2	grains enclosed by two papery 'glumes'	(traditional) oats, p. 12
3	ears with long straight awns	6-rowed barley, p. 9
2-4	fan-shaped spikelets with long awns	durum wheat, p. 7
3+	fan-shaped spikelets with no, or very short, awns	(beardless) wheat, p. 7
3+	fan-shaped spikelets with widely-spreading awns	bearded wheat, p. 7
3+	fan-shaped spikelets, short awns, very long ears	triticale, p. 11
3+	like oats but extra grains visible beneath 'glumes'	naked oats, p. 12

Maize, p. 13, has male flowers in a 'tassel' at the top and the cobs lower down the stem.

2. Broadleaved plants that flower in a normal growing season

If flowers are...	and the plant has...	it could be...
Red	a tall climbing growth habit & leaves with 3 large leaflets	runner bean, p. 40
	leaves with 3 small downy leaflets & many tiny flowers clustered into conical heads	crimson clover, p. 44
Orange	daisy-like flowers & mid-green leaves	calendula, p. 57
Yellow	*flowers with 4 identical petals:*	
	4mm flowers & arrow-shaped leaves	camelina, p. 35
	lobed hairy leaves & hairy seed capsules	white mustard, p. 25
	smooth upper leaves that clasp the stem	oilseed rape, p. 23
	smooth upper leaves with stalks	brown mustard, p. 24
	flowers that are daisy-like:	
	flowers 10-40cm across	sunflower, p. 56
	flowers up to 8cm across	Jerusalem artichoke, p. 62
	another type of flower:	
	leaves with 7-9 identical leaflets	yellow lupin, p. 43
	leaves with 3 small rounded leaflets, and a sprawling or straggling growth habit	trefoil, p. 44
	leaves with 3 minutely toothed leaflets, and an upright growth habit	sweet clover, p. 63
Green	leaves with 3-11 toothed leaflets	hemp, p. 58
	leaves that are not divided	quinoa, p. 63
Blue	*flowers with 5 identical petals:*	
	slender stem & narrow hairless leaves	linseed, p. 51
	hairy leaves & star-shaped flowers	borage, p. 59
	hairy stalk-less leaves & trumpet-like flowers	echium, p. 60
	flowers that are daisy-like:	
	dandelion-like leaves	'Stand & deliver', p. 62

If flowers are...	and the plant has...	it could be...
Blue (contd)	*another type of flower:*	
	leaves with 3 leaflets	lucerne, p. 45
	leaves with 7-11 identical narrow leaflets	blue lupin, p. 43
	leaves with 7-9 identical wide leaflets	white lupin, p. 43
Purple, mauve, lilac or pink	*flowers with 4 identical petals:*	
	grey leaves & flowers with large petals	culinary poppy, p. 61
	hairy leaves and pointed seed capsules	fodder radish, p. 32
	large flat, disc-like seed capsules	lunaria, p. 36
	flowers with 5 identical petals:	
	divided hairy leaves & pentagon-shaped flowers with large yellow stamens	potato, p. 50
	hairy stalk-less leaves & trumpet-like flowers	echium, p. 60
	fern-like hairy leaves and scented flowers	phacelia, p. 63
	heart- or arrow-shaped leaves	buckwheat, p. 62
	another type of flower:	
	leaves with 3 large pointed leaflets	dwarf bean, p. 39
	leaves with 3 large rounded hairy leaflets	soya bean, p. 41
	leaves with 3 small hairy leaflets & tiny purple flowers in a rounded head	red clover, p. 44
	leaves with 3 small hairless leaflets & tiny pink flowers in a rounded head	Alsike clover, p. 44
	hairless leaves with 2-3 pairs of leaflets	field bean, p. 38
	large leaf-like stipules and branching tendrils	peas, p. 37
	leaves with small toothed stipules and 4-8 pairs of small leaflets, ending in a branched tendril ...	tares, p. 47
	leaves with 6-12 pairs of leaflets, no tendrils	sainfoin, p. 46
White	*flowers with 4 identical petals:*	
	hairless leaves & stem, 4-5mm flowers	cress, p. 34
	hairy leaves and pointed seed capsules	fodder radish, p. 32
	spherical seed capsules	crambe, p. 33
	flowers with 5 identical petals:	
	divided hairy leaves & pentagon-shaped flowers with large yellow stamens	potato, p. 50
	hairy stalk-less leaves & trumpet-like flowers	echium, p. 60
	slender stem & narrow hairless leaves	linseed, p. 51
	another type of flower:	
	large leaf-like stipules and branching tendrils	peas, p. 37
	hairless leaves with 2-3 pairs of leaflets	field bean, p. 38
	a tall climbing growth habit & leaves with 3 large leaflets	runner bean, p.40
	leaves with 3 large pointed leaflets	dwarf bean, p. 39
	leaves with 3 large rounded hairy leaflets	soya bean, p. 41
	leaves with 3 small hairless leaflets, each with a white horseshoe-shaped mark	white clover, p. 44
	leaves with 7-9 identical wide leaflets	white lupin, p. 43
	leaves with 7-11 identical narrow leaflets	blue lupin, p. 43

Crops directory 1

Grasses and related species (monocotyledonous species)

This directory includes the most important plant family in agricultural systems worldwide: the grasses. These species provide cereal grains, and grazing for livestock. The identification of all flowering plants is based on flower structure, and is often not easy in grasses; the flowers and their constituent structures are often very small and the flower heads or ears are complex. However by looking at the building blocks of grass flowers, the spikelets, identification becomes much easier. Prior to flowering, grasses can be identified by looking at characteristics of the junction between leaf sheaths and leaf blades. In this way it is simple to tell the main cereals, and other grasses, apart.

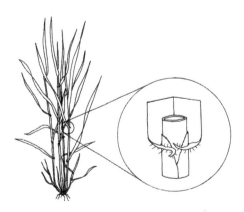

Look at the base of the leaf blade to identify grasses that are not in flower (in ear).

Wheat 1

1) Bread (or common) wheat

Scientific name: *Triticum aestivum* L. (syn. *T. vulgare* Host.).

Description: Annual cereal. Up to 1m tall, with two forms: beardless (most common form) and bearded (ears have spreading bristle-like awns: only one such variety currently grown in Britain). The grain is naked (has no hull), and there are up to 5 grains per fan-shaped spikelet.

Origin & history: SW Asia. Initially domesticated in the 7th millenium BC, after which many cultivated *Triticum* species were developed by hybridisation (both accidental and intentional) within the genus and with closely-related wild grass species. Wheats were introduced to Britain in the 4th millenium BC. *T. aestivum* is the most recently developed kind, originating south of the Caspian, and has replaced most of the other species in the last 2,000 years. Its grain is best suited to bread-making.

Use(s): The ripe seed (grain) is milled to produce various kinds of flour or used as animal feed, depending on the variety grown and the grains' protein content and quality. The grain is also used industrially as a source of starch, and has potential for bio-ethanol production (liquid fossil fuel substitute/additive). The straw is used for animal bedding, and that of specially-grown, obsolete, long-strawed varieties, is used for thatching.

Sown: Winter wheat (98% of crop) late Sep-end Oct. Spring wheat Feb-Apr.

Harvested: Winter wheat mid Aug-mid Sep. Spring wheat late Aug-mid Sep.

Grown: Countrywide, especially on heavier soils (cf. barley).

2) Durum (or macaroni) wheat

Scientific name: *Triticum turgidum* L. Durum group (syn. *T. durum* Desf.).

Description: Annual cereal. Similar to bread wheat, but the ears always have long awns extending far beyond the terminal spikelets. There are two to four naked grains per spikelet.

Origin & history: Common in the Mediterranean region, with traditional growing areas in Spain, Italy, Portugal and Greece. Durum was cultivated by the Ancient Egyptians, Greeks and Romans and first trialled in E England in the 1970s, but it remains a niche crop in Britain. The name derives from the Latin word for 'hard'.

Use(s): The ripe grain is milled to produce semolina for making pasta, or used in breakfast cereals.

Sown: Either Oct-early Nov, or Feb-mid Mar.

Harvested: Early Aug to preserve grain quality.

Grown: Mainly in Norfolk, Suffolk, Essex and Kent.

Wheat 2

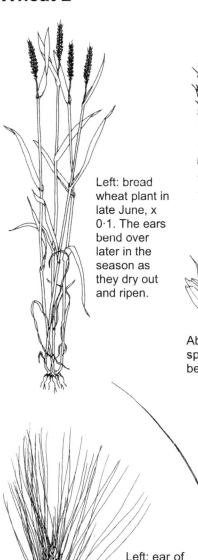

Left: bread wheat plant in late June, x 0·1. The ears bend over later in the season as they dry out and ripen.

Above: ears (x 0·5) and (below) fan-shaped spikelets (x 1) of beardless wheat (left) and bearded wheat (right).

Below: detail of leaf base of both species. Note the characteristic hairy 'auricles', (arrowed).

Left: ear of durum wheat, x 0·3. It has long awns on fan-shaped spikelets.

Right: durum spikelet, x 1.

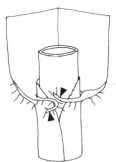

Barley 1

Scientific name: *Hordeum vulgare* L. (syn. *H. sativum* Jess., *H. distichum* L., *H. hexastichum* L.).

Description: Annual cereal. Up to 0·75-1m tall, with two forms: six-rowed and two-rowed. The ears have long awns. The majority of the crop is winter two-rowed varieties. All of the British barley crop has hulled grain, but naked barleys are also known.

Origin & history: SW Asia. Domesticated in the 7th millenium BC and introduced to Britain in the 4th millenium BC. Many naked primitive barleys were developed early in the crop's domestication, before naked wheats were known. Naked grain is more palatable and requires less processing than hulled grain, so naked barley was grown for human consumption in unleavened breads. Until naked wheats were developed, barley ousted wheat in Britain. 'Four-rowed' barley called bere (which is actually a lax form of the six-rowed type) was introduced to the North by the Vikings. Two-rowed barley was not grown in N Europe until the Middle Ages. The majority of the crop was only used by maltsters until the C20th, during which barley began to be extensively used as an animal feed. Naked barley is not currently grown in Britain, although one new variety is soon to be released. Novel 'waxy barley' (varieties that have different starch chemistry to traditional barley) are currently being developed by breeders.

Use(s): The ripe grain is used (i) to produce malt, the starting ingredient of beers and whisky (grain is steeped in water, allowed to germinate, then kilned) (two-rowed varieties only), (ii) as animal feed, or (iii) for processing into pearl barley and barley flour (the latter is used in barley water drinks and occasionally in baking), depending on the variety grown, its quality and the grains' nitrogen content. The straw is used for animal feed/bedding. A small crop of bere is still grown for milling to make flour for artisan bread-making. Waxy barley has potential uses as a fat-replacement agent in the food industry and also as an ingredient of breakfast cereals.

Sown: Winter barley mid Sep-mid Oct. Spring barley mid Jan-late Apr.

Harvested: Winter barley mid Jul-mid Aug. Spring barley last 3 wks Aug.

Grown: Countrywide, especially on lighter soils, with a large area in East Anglia and E Scotland. The best malting barley is said to come from Norfolk.

Barley 2

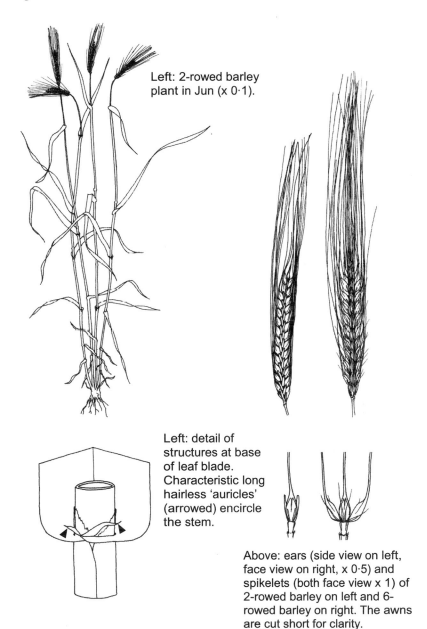

Left: 2-rowed barley plant in Jun (x 0·1).

Left: detail of structures at base of leaf blade. Characteristic long hairless 'auricles' (arrowed) encircle the stem.

Above: ears (side view on left, face view on right, x 0·5) and spikelets (both face view x 1) of 2-rowed barley on left and 6-rowed barley on right. The awns are cut short for clarity.

Rye

■ ☐ ☐ ☐ ☐

Scientific name: *Secale cereale* L.

Description: Annual cereal or fodder crop. Up to *c*. 1m tall, but crops contain individuals with a range of different heights. The leaves are greyish and the ears have short awns and 2 double rows of naked grain (2 per spikelet), the tips of which are visible between the floret structures.

Origin & history: SW Asia. A weed of primitive wheat crops. When cereals were first introduced to N Europe, it was impossible to separate the seed of the two species. Rye's good winter hardiness allowed it to out-compete primitive wheat and become a crop in its own right in the 1st millenium BC. It was first grown in Britain on a large scale in the Middle Ages and was once more important than wheat for bread-making in N Europe.

Use(s): The ripe seed of grain varieties is used for making crisp-breads. Young plants of forage varieties are used as temporary spring-time animal grazing, especially by dairy farmers. The straw is sometimes used for thatching.

Sown: Sep-mid Oct.

Harvested: Forage rye grazing finishes mid Apr. Grain crops cut mid Aug.

Grown: Grain crops on sandy land in S & SE England and E Anglia, where yields of other cereals would be poor.

Left: rye plant in early Jun, x 0·1.

Below: detail of structures at base of leaf blade, x 2.

Left: face view of ear, x 0·5.

Above: spikelet, x 1. Each contains 2 grains, giving rye ears their characteristic appearance.

Triticale

■■□□□

Scientific name: x *Triticosecale* Wittmack.

Description: Annual cereal or fodder crop. Up to 1m tall, with extremely long ears bearing short awns. The crop has naked grain, the tips of which are visible between floret structures. It is superficially similar to rye except that the spikelets are fan-shaped like wheat.

Origin & history: A man-made hybrid of durum wheat and rye, first developed at the end of C19th to combine the environmental tolerance of rye with the grain quality characteristics of wheat. Suitable varieties for British growers were only available from the 1970s. The grain has a high content of the essential amino acid lysine and the crop can out-yield wheat on poor soils.

Use(s): The ripe grain (winter varieties) is used as a 'high-lysine wheat' in pig and poultry food. Young plants are used like rye as an early grazing crop. The whole plant of spring varieties is used for whole-crop silage, either alone or with lupins. The crop is also included in some game cover mixtures.

Drilled: Winter triticale early-mid Oct.

Harvested: Forage triticale grazing finishes at the end of Apr. Grain crops are cut in Aug.

Grown: In similar situations to rye.

Left: plant in early Jul, x 0·1.

Right: ear, x 0·5. The awns on other varieties can be slightly longer than these.

Above: spikelet, x 1. It is fan-shaped, like its wheat parent.

11

Millet (Proso millet)

■☐☐☐☐

Scientific name: *Panicum miliaceum* L.

Description: Annual cereal or fodder crop. Up to 1m tall, with hairy leaf sheaths and large drooping seed heads (panicles), containing white or red seeds. The crop is fast maturing and drought tolerant.

Origin & history: C Asia. Cultivated in Bronze-Age Europe, in some regions before wheat was introduced. Today millet is mostly grown in C Asia, India and the Middle East. It was given the name 'milium' by the Romans because of large numbers of grain in each panicle.

Use(s): The ripe grain is used as an ingredient of bird food mixes (but can also be used in speciality breads etc.) and the straw is used for animal feed. The whole plant is harvested whilst still green as whole-crop silage for animal feed. Whole plants are also grown to maturity as game cover.

Sown: Late Apr-early May.

Harvested: Whole-crop silage in Aug. Grain crop late Aug-early Sep.

Grown: S England for grain, elsewhere for game cover.

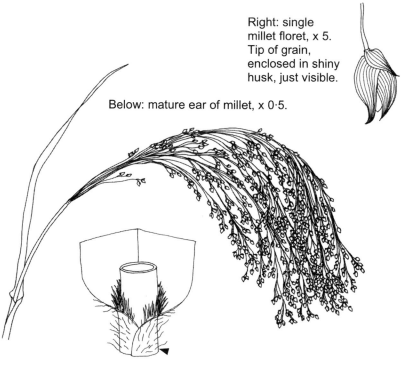

Right: single millet floret, x 5. Tip of grain, enclosed in shiny husk, just visible.

Below: mature ear of millet, x 0·5.

Above: detail of leaf base structures, x 1. Note the hairy leaf sheath (arrowed).

Miscanthus ('elephant grass')

Scientific name: *Miscanthus* x *giganteus* Greef & Deuter ex Hodkinson & Renvoize. Often called 'elephant grass', though this name strictly applies to the African grass species *Pennisetum purpureum* Schum.

Description: Perennial energy and fibre crop. Up to 3·5m tall. Dies down each winter leaving sturdy cane-like stems.

Origin & history: The genus was introduced into Europe in the C19th as garden ornamentals. *Miscanthus* x *giganteus* is a sterile hybrid originating in Japan and is relatively new to Europe. Established crops last at least 20 years. The crop area is likely to continue increasing as the demand for renewable energy rises.

Use(s): The mature, dry canes are used as a source of biomass, primarily for fuelling power stations, but also for equine bedding & cat litter, bio-degradable plant pots, constructional fibreboards and are under evaluation as a novel thatching material. Miscanthus is additionally planted as a screen, mini-shelter belt, or as game cover.

Planted: As rhizomes in Mar & Apr.

Harvested: Jan-Mar.

Grown: Large area in Somerset. The most northerly commercial crop is in Lincs.

Left: part of mature crop in mid summer, x 0·05.

Right: detail of leaf blade base on lower leaf, x 0·5.

Above right: rhizome in Mar, x 0·3.

Other species, including reed canary grass (p. 62) and switch grass (*Panicum virgatum* L.), are also currently under evaluation for energy production.

15

Herbage grass seed crops

Scientific names: Several species are grown, but the four most common are perennial ryegrass (PRG; *Lolium perenne* L.), Italian ryegrass (IRG; *Lolium multiflorum* Lam.), cocksfoot (*Dactylis glomerata* L.) and timothy (*Phleum pratense* L.).

Description: Perennial grasses. Italian ryegrass is shorter lived than the others.

Origin & history: Perennial ryegrass and cocksfoot are native species. Cocksfoot has a long history of use on drought-prone soils. Italian ryegrass originated in N Italy, and was introduced in the C19th. It is faster growing than perennial ryegrass. The origin of timothy is unclear; it was first recognised as a good grazing grass in New Hampshire, USA, then introduced to Maryland by a farmer called Timothy Hansen in *c.* 1720, and finally brought to Britain in 1760. All the species have been improved by breeding work beginning in the C20th.

Use(s): All the species are used for animal grazing, but small areas are grown as arable crops for seed production for future sowings.

Sown: The ryegrasses are either undersown into a spring cereal or drilled alone. Cocksfoot and timothy are drilled alone from spring to July.

Harvested: Crops are swathed from mid Jul-mid Sep depending on species, then combined.

Grown: Mainly in East Anglia.

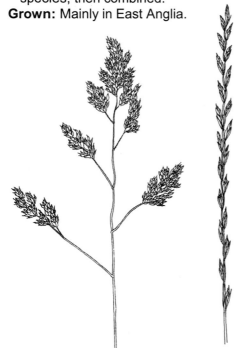

From far left: ears (x 0·5) of cocksfoot, PRG and timothy. PRG and IRG have different spikelet types (below).

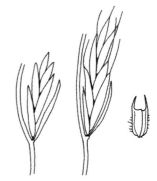

Above from left: spikelets of PRG, IRG (which is larger than PRG and also has short awns) and timothy, all x 3.

Onion and shallot

Scientific name: *Allium cepa* L. Cepa group (onion) and Aggregatum group (shallot, syn. *A. ascalonicum* L.).

Description: Onion: a biennial usually grown as an annual vegetable crop. Onion leaves consist of two parts: the lower part is thickened to form the familiar onion bulb, whilst the upper part is thin and hollow (and not present in dry onions offered for sale). Shallot: a form in which the axillary buds develop into a clump of small bulbs. Shallots rarely flower, so are propagated by division of clumps.

Origin & history: Turkistan. Long in cultivation and a staple in the Ancient Egyptians' diet. Six different types were mentioned by Pliny in the C1st AD in Rome. The crops are grown worldwide and are an essential ingredient in the cuisine of all countries. Over-wintering onions were derived from Japanese varieties. Shallots are especially used in French cooking.

Use(s): Vegetables for human consumption.

Sown: Over-wintering crops in early-mid Aug. Main crop in Mar & Apr. Some crops are established from sets (immature small bulbs grown the preceding year), or from transplants originally sown in Jan & Feb.

Harvested: Late Aug-Sep, after foliage bends over.

Grown: Mainly in the E counties of England, and in Lincs., Beds., Yorks. and Kent.

Young onion plant, x 0·3.

Leek

Scientific name: *Allium ampeloprasum* L. Porrum group (syn. *A. porrum* L.).

Description: Biennial grown as an annual vegetable crop. It has broad, flat leaves, V-shaped in cross section. The thickened leaf sheaths form a cylindrical shank (shoot base).

Origin & history: Mediterranean. Widely cultivated by the Middle Ages in N Europe because of its better cold tolerance than onions and long harvesting period (since its growth is continuous and no bulb is formed). Its popularity waned in the C16th & 17th. The leek is the national emblem of Wales, commemorating a military victory over the Saxons in AD 640. Northumbrian gardeners grow especially large, competition leeks.

Use(s): A vegetable for human consumption.

Sown: In modules Jan-Apr, and planted out Apr-Jul. Seed is also direct sown in Mar & Apr.

Harvested: July and over winter to following May.

Grown: Large crop in the Fens.

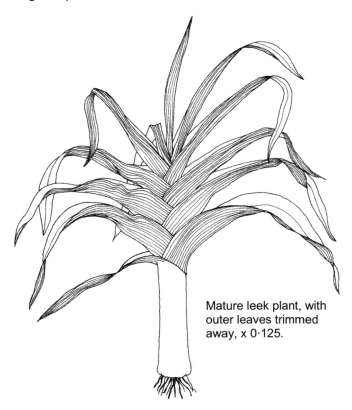

Mature leek plant, with outer leaves trimmed away, x 0·125.

Garlic

■□□□□

Scientific name: *Allium sativum* L.

Description: Perennial grown as an annual vegetable crop. It has flat leaves, and a compound white or purple-skinned bulb, split into cloves with a characteristic pungent smell. There are two forms: hardneck, with a persistent 'flower' stem, and softneck, without such a stem. A small area of the much larger crop, elephant garlic (*A. ampeloprasum* Ampeloprasum group), is also grown. This is similar to garlic, but it can reach 1·5m in height and has bigger bulbs.

Origin & history: In cultivation since the 4th millenium BC around the Mediterranean, and was possibly derived from a wild central Asian species. World-wide, garlic is the second most important *Allium* crop after onions. It has a long history of unpopularity in Britain (was once called 'stinking rose'), and this has only reversed during the last few decades with increasing interest in Mediterranean food. Field scale crops were introduced around 25 years ago in the Isle of Wight.

Use(s): The swollen bulb and its constituent cloves are used as a vegetable/herb.

Planted: Oct-Feb.

Harvested: May-Jul.

Grown: On well-drained soils. Large crop in the Isle of Wight. The most northerly commercial grower is in Scotland.

Garlic plant
in late Jun, x 0·3.

Asparagus

■☐☐☐☐

Scientific name: *Asparagus officinalis* L.

Description: Perennial vegetable crop, grown in beds. Up to 1·5m tall, with ferny foliage. The asparagus fern of florists is the related, but inedible, ornamental species *A. setaceus* (Kunth) Jessop.

Origin & history: S Europe & N Africa. Asparagus has been in cultivation since the time of the Ancient Greeks, but was abandoned in the Middle Ages, except by the Arabs. The crop then became fashionable once again in C17th France. The subspecies *prostratus* is a rare British seaside native. Wild asparagus is still gathered in S Europe.

Use(s): The newly-emerged shoots ('spears') are a gourmet vegetable.

Planted: 'Crowns' (fleshy rootstocks) of all-male varieties are planted in spring. The crop lasts around 15 years.

Harvested: Late Apr-early Jun. Uncut shoots are allowed to grow, then die off and are cut down in Feb.

Grown: S England, usually on the very lightest sandy soils under irrigation. Traditional areas for production include the Vale of Evesham and the Fens.

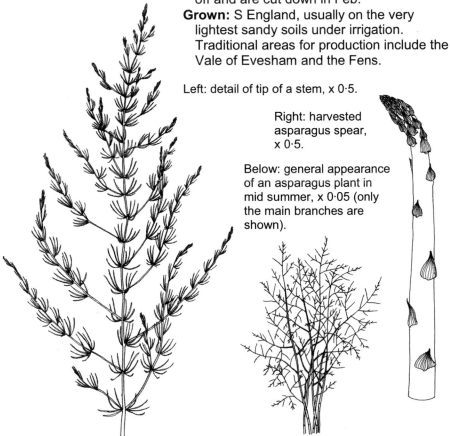

Left: detail of tip of a stem, x 0·5.

Right: harvested asparagus spear, x 0·5.

Below: general appearance of an asparagus plant in mid summer, x 0·05 (only the main branches are shown).

Crops directory 2

Broadleaved plants
(dicotyledonous species)

This directory contains plants in ten dicotyledonous families. Of these, the family currently occupying the largest area in Britain is the cabbage family (Cruciferae or Brassicaceae). Its members are easily recognised by their flowers, which have 4 petals arranged in a cross shape and almost always 6 stamens (e.g. oilseed rape, p. 23). Many, but not all, of these plants also have a 'pinnatifid' leaf shape (typified by the oilseed rape leaf, also on p. 23).

The second most widely planted dicotyledonous family is the pea family (Leguminosae or Fabaceae). Its members (called legumes) have root nodules containing nitrogen-fixing bacteria, which can convert atmospheric nitrogen into nitrate (nitrate is an important plant nutrient). Legumes are vitally important agricultural plants, because their cultivation increases soil fertility and they have a high protein content, making them valuable as animal and human food. Legume flowers mostly consist of 5 petals in a characteristic arrangement: the upright 'standard', 2 'wing' petals and a 'keel' comprising 2 more petals which can be fused along part of their length. A good example is the pea flower on p. 37. Legume leaves are often made of leaflets e.g. clover leaves comprise 3 leaflets.

Rape crops 1

Scientific name: *Brassica napus* L.

1) Oilseed rape (OSR, canola or coleseed)

Description: Annual winter- or spring-sown oilseed crop. Up to 1-1·5m tall, with hairless, waxy adult leaves (young leaves slightly hairy) and pungent yellow flowers followed by smooth, narrow, cylindrical seed capsules.

Origin & history: Mediterranean. All *Brassica* species have been long in cultivation. Oilseed rape was probably developed in the Middle Ages for lamp oil. Only a small amount was grown in Britain until the 1960s/70s. World-wide it now accounts for 13% of oilseed production. Similar forms were developed in *B. rapa* (turnip rapes), but are now obsolete in Britain.

Use(s): The ripe seed is crushed for oil: (i) from double-low rape (low in erucic acid and glucosinolates) used for human consumption as cooking oils, in margarine etc. and (ii) from high erucic acid rape (HEAR), used for industrial purposes, lubricants, biodiesel etc. Rapeseed meal (by-product of crushing) is a component of animal feeds. Rape seed is sometimes used as substitute for white mustard in 'mustard & cress' salad punnets.

Sown: Winter rape (the majority of the crop) mid Aug-mid Sep. Spring rape late Mar-mid Apr.

Harvested: Winter rape late Jul-mid Aug. Spring rape early Sep. The seeds of winter rape ripen unevenly, so crops are usually swathed (stems cut and plants left to dry out on the field) or desiccated (treated with a herbicide to kill the plants and allow them to dry out) before combine-harvesting. Spring rape is often directly combined.

Grown: Countrywide.

2) Forage rape & rape kales

Description: Annual forage crops. Forage rape is up to *c.* 0·7m tall.

Origin & history: Widely grown in Britain before oilseed rape. Other kinds of forage rapes were developed in the 1970s from crosses of *B. rapa* with an oriental salad brassica.

Use(s): Whole plants of forage rape varieties are used as animal feed or game cover. 'Hungry gap' and rape kale are mostly grazed by sheep in the spring. Rape kale is sometimes included in game cover mixtures.

Sown: Forage rape mid Apr-mid Aug. 'Hungry Gap' and rape kale drilled into cereal stubble in autumn.

Harvested: Forage rape is usually fed *in situ* to sheep around 10 weeks after sowing. 'Hungry Gap' and rape kale is used in Mar & Apr.

Grown: Forage rape does well in the W, N and in Wales. 'Hungry Gap' and rape kale are only grown in the S of the country.

Rape crops 2

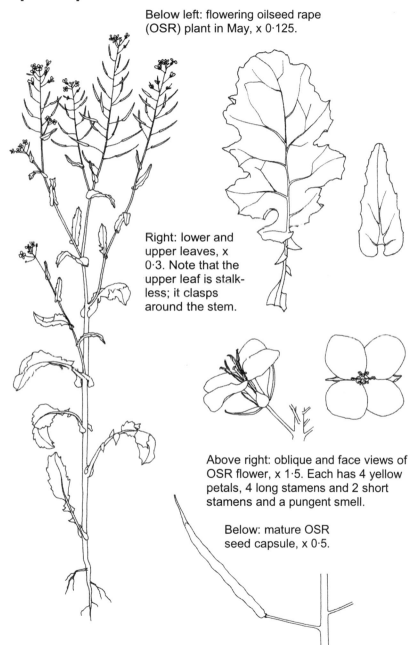

Below left: flowering oilseed rape (OSR) plant in May, x 0·125.

Right: lower and upper leaves, x 0·3. Note that the upper leaf is stalkless; it clasps around the stem.

Above right: oblique and face views of OSR flower, x 1·5. Each has 4 yellow petals, 4 long stamens and 2 short stamens and a pungent smell.

Below: mature OSR seed capsule, x 0·5.

Brown (Indian or Chinese) mustard ■☐☐☐☐

Scientific name: *Brassica juncea* (L.) Czerniak.

Description: Annual seed crop. Up to 1·5m tall, with yellow flowers. Similar to oilseed rape, but flowers and seed capsules are smaller, upper leaves have distinct stalks and brown mustard is in flower later in the season.

Origin & history: NW India. Grown as an oilseed or a leafy green vegetable in Asia. The crop was introduced to Britain in around 1952 to replace black mustard (*B. nigra* L., syn. *Sinapis nigra* L.), a native species which is less compact, more prone to seed shedding, and is difficult to harvest mechanically.

Use(s): The ripe seed is an ingredient of table mustard (it gives pungency).

Sown: Late Feb-late Apr.

Harvested: Late Aug.

Grown: E counties of England on contract to Colman's of Norwich.

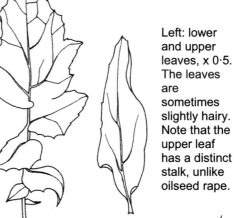

Left: lower and upper leaves, x 0·5. The leaves are sometimes slightly hairy. Note that the upper leaf has a distinct stalk, unlike oilseed rape.

Right: flowering plant in early Jul, x 0·1. The flowers are similar to oilseed rape: yellow, with 4 petals.

Right: mature seed capsules, x 0·5.

White mustard

■□□□□

Scientific name: *Sinapis alba* L. (syn. *Brassica hirta* Moench).

Description: Annual seed, forage or green manure crop. Up to 1·3m tall, sparsely hairy and rough-stemmed, with yellow flowers followed by bristly seed capsules.

Origin & history: Mediterranean and Crimea. In cultivation since Biblical times and introduced to Britain by the Romans.

Use(s): The ripe seed is an ingredient of table mustard (it gives heat), or is germinated for salad 'mustard & cress'. Whole plants are used as a green manure to conserve soil nutrients, as a temporary autumn forage crop, or as game cover.

Sown: Condiment mustard late Feb late Apr; otherwise, in the autumn.

Harvested: Condiment mustard late Aug.

Grown: Condiment use, see brown mustard. Other uses, countrywide.

Left: flowering plant in mid Jun, x 0·1. The stem is rough and bristly, not smooth and waxy. The flowers (*c.* 10mm across) have 4 yellow petals.

Right: lower leaf, x 0·25. Leaves hairy.

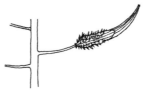

Above: mature seed capsule, x 0·5. It is bristly, with a long flat 'beak' at far end, and is held at about 90° to the stem.

Brassicas 1

General origin & history: Brassica crops are derived from the perennial wild sea cabbage of S & W Europe and have been in cultivation for around 4,000 years. The species has diverged into a number of very distinct separate crops. All have waxy hairless leaves and are biennials.

1) Cabbage

Scientific name: *Brassica oleracea* L. Capitata group.
Description: Has a tightly-packed head comprising a massively enlarged terminal bud; rounded in winter cabbages, pointed in spring cabbages.
Origin & history: Cabbages probably developed in Germany in the C12th. Several different groups are grown as vegetables: spring cabbages (spring greens), summer & autumn cabbages, winter cabbages (including red cabbage and savoys, the latter introduced from Italy in the C16th). Large cattle cabbages were formerly an important fodder crop, but are now only grown on small farms.
Use(s): The heads are used as a vegetable for human consumption, or (different varieties) for cattle feed.
Sown: Throughout the year depending on type. Either sown in modules or nursery beds then transplanted, or direct sown.
Harvested: Throughout the year depending on maturity.
Grown: Spring cabbage in SW England, Kent and Lincs. Others countrywide on chalky soils.

2) Cauliflower

Scientific name: *B. oleracea* Botrytis group.
Description: Has cabbage-like leaves surrounding a dense white mass of immature flower buds (the 'curd'). Novelty forms with purple or green curds are grown in gardens. 'Romanesco' is a type of green cauliflower.
Origin & history: E Mediterranean. The crop was developed by the Arabs in the Middle Ages, introduced to Italy in the Renaissance, and to NW Europe in the C17th.
Use(s): The curd is used as a vegetable.
Sown: Feb-early Jun in modules or nursery beds for transplanting Mar-Jul.
Harvested: Summer crops Jul-Sep, autumn crops Sep-Dec, winter-heading crops Dec-May and over-wintering crops in May & Jun.
Grown: Most of the crop is in S Lincs. Some is also grown in W Cornwall (winter-heading types), S Pembrokeshire, the Isle of Thanet, around Evesham and in Lancs.

Brassicas 2

3) Broccoli and calabrese

Scientific name: *Brassica oleracea* L. Italica group.

Description: Axillary and terminal buds form branching flower stalks.

Origin & history: Developed in Italy, especially around Calabria. It was introduced to England as a gourmet vegetable in the early C18th.

Use(s): The immature flowering stalks or buds are used as a vegetable for human consumption.

Sown: Feb-late Mar under glass for transplanting early-mid Apr. Alternatively direct drilled mid Mar-mid Aug.

Harvested: Late May-late Nov.

Grown: Mainly in the Fens.

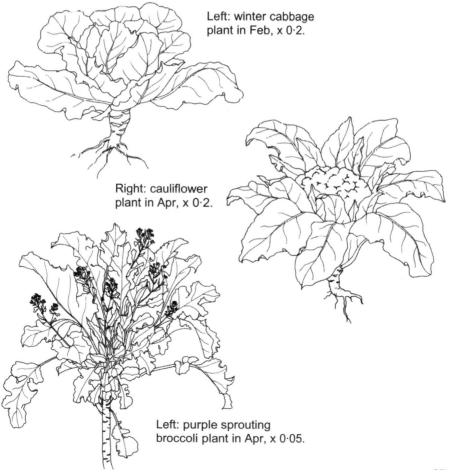

Left: winter cabbage plant in Feb, x 0·2.

Right: cauliflower plant in Apr, x 0·2.

Left: purple sprouting broccoli plant in Apr, x 0·05.

Brassicas 3

3) Brussels sprouts   ■□□□□

Scientific name: *Brassica oleracea* L. Gemmifera group.
Description: Up to 1m tall, with a loose, cabbage-like head and enlarged axillary buds along a thick, woody stem.
Origin & history: The first certain records come from C18th Belgium.
Use(s): The axillary buds are used as a vegetable for human consumption. The stems ('sprout stalks') are used for cattle feed.
Sown: Mid Feb-mid Apr and transplanted May & early Jun, or direct drilled.
Harvested: Late Aug-early Mar. Either picked over several times to remove sprouts as they mature, or (in the majority of crops) harvested in a single operation by specialised machinery.
Grown: Widely on fertile soils.

4) Kohl-rabi   ■□□□□

Scientific name: *Brassica oleracea* L. Gongylodes group.
Description: Low growing plant with thickened stem which forms a globular 'bulb' with green or purple skin.
Origin & history: Introduced from Germany in the early C19th.
Use(s): Either a vegetable for human consumption, or for animal feed.
Sown: Mid Mar-Jul.
Harvested: 9-12 weeks after sowing.
Grown: Large crop on limestone soils in Lincs. Kohl-rabi can succeed on shallow soils where turnips and swedes would fail.

5) Kale   ■□□□□

Scientific name: *Brassica oleracea* L. Acephala group.
Description: Up to 1m tall, kale is a leafy plant which occurs in two main types: (i) marrowstem, with a thick stem up to 10cm in diameter and (ii) thousand-head, with a smaller stem, but more numerous leafy side shoots. There are also hybrids.
Origin & history: Cultivated for at least 4,000 years. This crop is the closest to the appearance of the wild sea cabbage. Marrowstem kale was first recorded in France in the early C19th. Kales are also grown as garden vegetables. 'Hungry Gap' and rape kale are types of *B. napus* (p. 22).
Use(s): Mostly grown as a fodder crop, but also planted for game cover.
Sown: Apr-Jun.
Harvested: Marrowstem is used before Christmas (mostly for dairy cows), thousand-head is used Christmas-Mar (mostly for sheep).

Brassicas 4

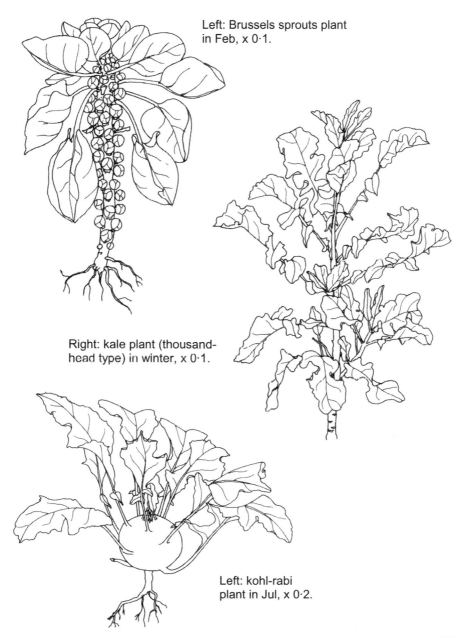

Left: Brussels sprouts plant in Feb, x 0·1.

Right: kale plant (thousand-head type) in winter, x 0·1.

Left: kohl-rabi plant in Jul, x 0·2.

Turnip ■■□□□

Scientific name: *Brassica rapa* L. Rapifera group.

Description: Biennial usually grown as an annual root or forage/fodder crop. It has sparsely hairy leaves, a swollen root with white or yellow flesh and white skin (sometimes green or purple at the top of root).

Origin & history: European. Cultivated for *c.* 4,000 years. The wild form, locally common near rivers in Britain, has a slender taproot. Different types were developed: culinary turnips (the oldest use), forage turnips for roots (a storable winter feed for livestock, introduced and popularised by 'agricultural improvers', such as Turnip Townshend, in the C18th), turnip rapes (formerly used in Britain like oilseed rape, but now obsolete) and 'stubble turnips' (quick growing forms which are always white-rooted).

Use(s): The root is used as a vegetable, or animal feed. Whole plants of stubble turnips are used as a temporary autumn grazing crop (especially for sheep) and also in game cover mixtures.

Sown: Culinary turnips: earlies Mar-Jun, maincrop mid Jul-mid Aug. Forage turnips May-end Jun. Stubble turnips Apr-Aug.

Harvested: Culinary turnips: earlies May-Sep, maincrop mid Oct onwards. Forage turnips used Aug-Feb. Stubble turnips used 8-12 weeks after sowing.

Grown: Countrywide.

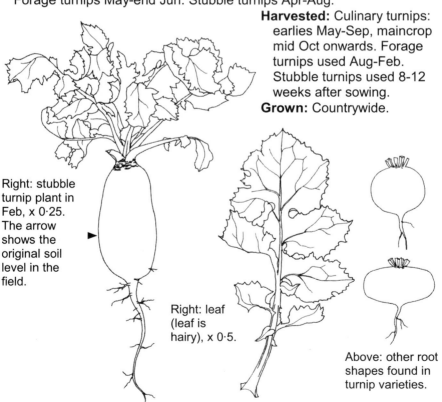

Right: stubble turnip plant in Feb, x 0·25. The arrow shows the original soil level in the field.

Right: leaf (leaf is hairy), x 0·5.

Above: other root shapes found in turnip varieties.

Swede (rutabaga in USA)

Scientific name: *Brassica napus* L. Napobrassica group.

Description: Biennial usually grown as an annual root crop. It has a swollen root with orange/yellow flesh and purple, bronze or green skin. Unlike turnips, swedes have a 'neck' between their swollen root and the leaves, on adult swede plants, are hairless.

Origin & history: European. First developed in C17th Bohemia and introduced to Britain via Sweden in the 1780s. Swedes are widely used, like turnips, to feed livestock over winter. The name 'rutabaga' is an anglicised form of the Swedish dialect name 'rotbagga', meaning ram's root. Rapes are also members of this species (p. 22-23).

Use(s): The roots are either used as a vegetable for human consumption (only those with purple or green skins), or for animal feed.

Sown: Culinary swedes Mar-Jul, or transplanted in Mar from Jan sowings. Fodder swedes Apr-Jun.

Harvested: Culinary swedes late Sep onwards. Fodder swedes used Nov-Mar (but green-skinned swedes can be stored till May in Scotland).

Grown: Most culinary swedes in Devon & Scotland. Green-skinned swedes mostly only grown in Scotland.

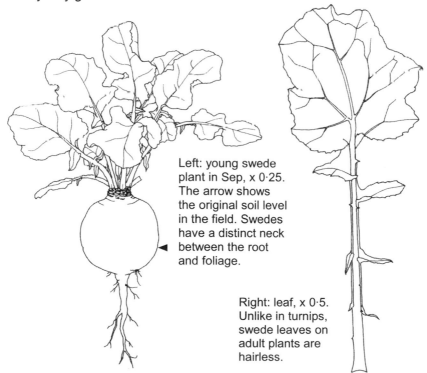

Left: young swede plant in Sep, x 0·25. The arrow shows the original soil level in the field. Swedes have a distinct neck between the root and foliage.

Right: leaf, x 0·5. Unlike in turnips, swede leaves on adult plants are hairless.

31

Fodder radish

■☐☐☐☐

Scientific name: *Raphanus sativus* L.

Description: Annual or biennial, usually grown as an annual forage crop. It has hairy leaves, a white-fleshed enlarged taproot, pink/lilac or white flowers and large seed capsules.

Origin & history: Grown by the Assyrians and also recorded in the 3rd millenium BC in Egypt. The crop quickly spread to the Orient, where large-rooted vegetable forms were developed. In Europe, radish is widely used as a salad vegetable. Fodder radishes were introduced to Britain in the 1960s. They are not frost hardy, but will grow on land infested with clubroot disease, where other crucifer species fail. Other radishes (not British field crops) are grown as oilseeds, or for their extremely large seed capsules that are harvested whilst immature for pickling (rat's tail radishes).

Use(s): Thickly sown crops produce a mass of foliage, which is grazed *in situ* as a temporary autumn fodder crop. Fodder radish is also included in game cover mixes, where it is allowed to flower and set seed.

Sown: Jun & Jul.

Harvested: Fodder crops are grazed within 6-11 weeks after sowing, otherwise the plants run to seed and become unpalatable.

Grown: Countrywide.

Left: one-month-old plant, x 0·25, with the root beginning to swell.

Right: ripe seed capsule, x 0·5 (from a plant in a game cover plot).

Far right: leaf, x 0·5. Both stalk and leaf blade are hairy.

Crambe (Abyssinian mustard/kale) ■□□□□

Scientific name: *Crambe abyssinica* Hochst. ex R. E. Fries.

Description: Annual industrial oilseed crop. It has mustard-like leaves and white flowers, followed by spherical seed capsules.

Origin & history: Turko-Iranian region of SW Asia. Crambe has only been known in the UK for around 10 years, and was not fully commercialised until 2001. It is either very closely related to, or synonymous with *C. hispanica* L. Experimental hybrids between *C. abyssinica* and *C. hispanica* have been made.

Use(s): The ripe seed is crushed for its inedible oil which is high in erucic acid (used especially in the plastics industry after conversion to erucamide). Crambe seed meal (residue of crushing) is a high-protein ingredient in some animal feeds. Crambe is said to have several agronomic advantages over HEAR (p. 22), the main alternative source of erucic acid.

Sown: Mid Apr-mid May.

Harvested: Late Jul-end Aug.

Grown: Can be grown in all arable areas.

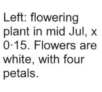

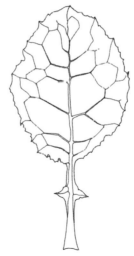

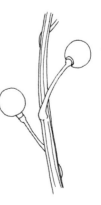

Left: leaf, x 0·5. Leaves are shiny with a crinkled surface and prominent veins.

Left: flowering plant in mid Jul, x 0·15. Flowers are white, with four petals.

Right: mature seed capsules, x 1. Easily recognisable because of their spherical shape, unique amongst British field crops.

Cress

■☐☐☐☐

Scientific name: *Lepidium sativum* L.

Description: Annual seed crop. Up to 0·7m tall, hairless, with divided leaves and white flowers followed by flat seed capsules. It has a characteristic smell.

Origin & history: SW Asia. According to the Ancient Greeks, cress was eaten by the Persians before bread was known. Cress probably spread as a secondary crop associated with flax (and was grown with flax in Moorish Spain). The mature foliage was formerly used as a vegetable, but this is not usually done nowadays in Britain.

Use(s): The ripe seed is germinated in punnets to the cotyledon stage for use as a salad.

Sown: Early Apr.

Harvested: Late Jul.

Grown: Small crop in Norfolk, Suffolk and Essex.

Left: flowering plant in Jun, x 0·2. The stem is thin, wiry, smooth, covered in greyish wax.

Above: leaves, all x 0·5. They are smooth, variable in shape (this is controlled by an incompletely dominant gene). The upper leaves are simple, not divided.

Above: flowers, x 1. Each has 4 white petals, 4 long stamens and 2 short stamens with brown tips. There is no scent.

Right: seed capsules (face and side views), x 4. The capsules are buff in colour and each contains 2 seeds.

Camelina (gold-of-pleasure)

■☐☐☐☐

Scientific name: *Camelina sativa* (L.) Crantz.

Description: Annual novel oilseed crop. Up to 0·7m tall, with arrow-shaped leaves and small yellow flowers, followed by egg-shaped seed capsules.

Origin & history: Mediterranean. In cultivation since the Bronze Age, and historically used to produce lamp oil. Formerly it was a specific weed of linseed crops. It has only recently been revived as an oilseed crop.

Use(s): The ripe seed is crushed for oil, rich in alpha-linolenic acid, linoleic acid and natural anti-oxidants (tocopherols). It is used as a culinary oil (has a pleasant almond-like aroma), and is thought to reduce cholesterol levels. Future industrial uses of the oil could be in skin care products, soaps and as a marine oil. Camelina is also sown in game/conservation mixes and is especially attractive to partridges.

Sown: Late Apr.

Harvested: Late Jul & Aug.

Grown: Can be grown in all arable areas.

Left: flowering plant in Jul, x 0·2. Note that specimens planted in closer rows may be less branched than this.

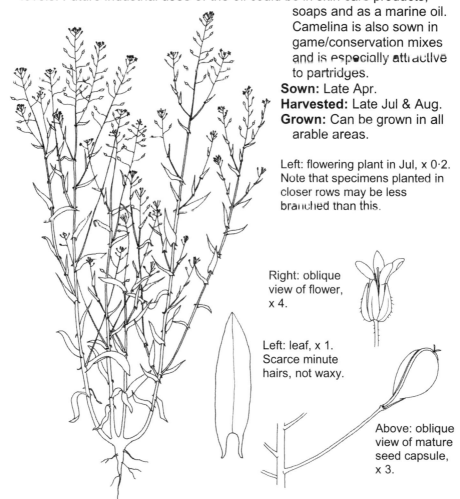

Right: oblique view of flower, x 4.

Left: leaf, x 1. Scarce minute hairs, not waxy.

Above: oblique view of mature seed capsule, x 3.

Lunaria (honesty)

■□□□□

Scientific name: *Lunaria annua* L.

Description: Biennial novel industrial oilseed. Up to 1m tall, with shiny, hairy leaves and purple flowers followed by thin, flat, oval seed capsules.

Origin & history: SE Europe & W Asia, but now naturalised in temperate Europe and N America. Lunaria was taken into cultivation as an ornamental, because of its decorative dried seed capsules. It was recently evaluated as a novel oilseed crop, and commercialisation aims to develop new annual varieties.

Use(s): The ripe seed is crushed for oil, rich in erucic acid and nervonic acid. The oil has two potential uses: (i) production of high-temperature lubricants and engineering nylons, (ii) in pharmaceutical preparations for the treatment of multiple sclerosis. The seed meal by-product of crushing is under evaluation for use in animal feeds.

Sown: Autumn.

Harvested: Aug.

Grown: Can be grown in all arable areas.

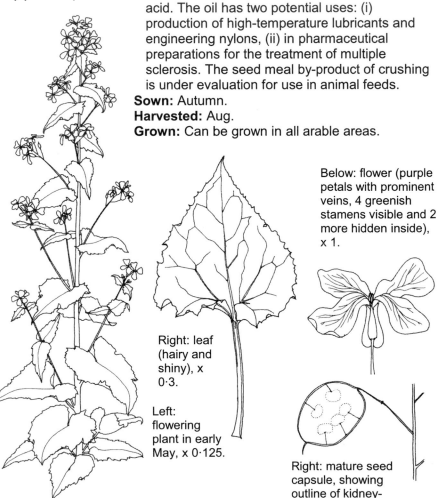

Below: flower (purple petals with prominent veins, 4 greenish stamens visible and 2 more hidden inside), x 1.

Right: leaf (hairy and shiny), x 0·3.

Left: flowering plant in early May, x 0·125.

Right: mature seed capsule, showing outline of kidney-shaped seeds, x 0·5.

Peas

Scientific name: *Pisum sativum* L. (syn. *P. arvense* L.).

Description: Annual pulse, vegetable or forage crop. Field crops are low-growing, with 4 leaf types (all with tendrils) and white or purple flowers.

Origin & history: Georgia (former Soviet Republic). Peas have been grown since prehistoric times over Europe. They were originally harvested as a dry pulse, but the C17th fashion for luxury immature seeds gave rise to their most familiar use as a vegetable. Forage peas were introduced to Britain in 1975.

Use(s): The immature seeds are used as a vegetable, either as picked peas (sent to market still in the pods), or as vining peas (mechanically harvested and quick-frozen or tinned). Immature whole pods of mange tout and sugar snap peas are used as a vegetable. Dry, mature seed of 'combinable' peas is used for human consumption (high quality seed from white-flowered peas only), or as a component of animal feed (including maple peas for pigeon feed). Whole plants of forage peas are used for animal feed.

Sown: Mid Feb-early Mar.

Harvested: Jul (earliest crops in S) to late Aug-early Sep (in N).

Grown: Countrywide, especially in low rainfall areas. Vining peas in E England near processing factories.

Right: flower, x 2. Petals have prominent green veins.

Above: pea plant (semi-leafless type), in early Jul, x 0·3.

Leaf types from left: normal, tare-leaved and semi-leafless. Leafless peas have tiny stipules and no leaflets.

Field beans (inc. broad and tic beans)

Scientific name: *Vicia faba* L.

Description: Annual pulse or vegetable crop. Up to 1·7m tall, hairless, multi-stemmed, with greyish leaves usually with 2-3 pairs of leaflets, fragrant black & white (or lilac) flowers in leaf axils, followed by stout seed pods. The whole plant turns dark brown on maturity.

Origin & history: Europe. Cultivated since Prehistoric times, originally for storage as a dried pulse. It is classified into horse beans (large, flat, oval seed) and tic/tick beans (smaller rounded seeds; always spring-sown). Broad beans are varieties developed for tender immature seeds.

Use(s): The immature seeds of broad beans are used as a vegetable. The mature, dry seed of others is used for animal feed. Tic beans are used especially in pigeon feed, and also as game cover crops.

Sown: Winter beans, late Oct-early Dec. Spring & broad beans late Feb-early Mar.

Harvested: Broad beans Jun-Sep. Other types in Sep.

Grown: Countrywide.

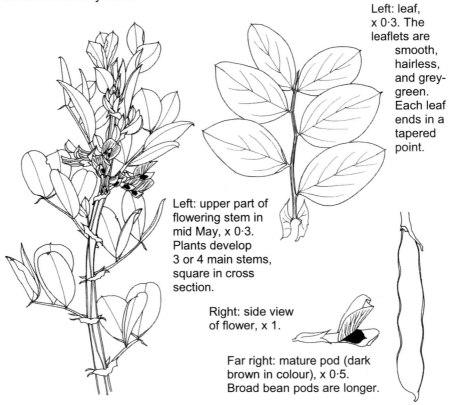

Left: leaf, x 0·3. The leaflets are smooth, hairless, and grey-green. Each leaf ends in a tapered point.

Left: upper part of flowering stem in mid May, x 0·3. Plants develop 3 or 4 main stems, square in cross section.

Right: side view of flower, x 1.

Far right: mature pod (dark brown in colour), x 0·5. Broad bean pods are longer.

Dwarf (French) beans

■☐☐☐☐

Scientific name: *Phaseolus vulgaris* L.

Description: Annual vegetable crop. Bush varieties are grown in fields; climbing varieties in gardens only. The leaves have 3 leaflets; the central one being larger than the other two. The flowers are mauve or white and are followed by green, yellow or purple pods.

Origin & history: C America. Cultivated in Mexico for 5,000 years and introduced to Europe by Spanish explorers in the C16th and then to Britain in 1594. It is commonly grown in gardens. There is a vast crop of navy beans (with small rounded white seeds) in the USA, some of which is imported to Britain to make tinned baked beans. Navy bean varieties have been trialled in Britain, but are not grown commercially.

Use(s): The unripe pods are used as a vegetable. The ripe, dry seeds (haricot beans) are used as a pulse for human consumption.

Sown: Early May.
Harvested: Jul-Sep.
Grown: SE England, or further N on S-facing slopes only. A small area in S England is suitable for haricot production.

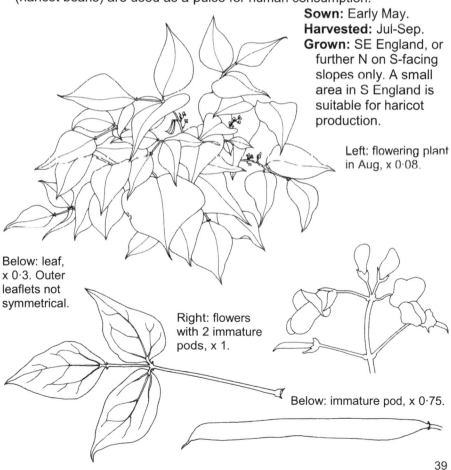

Left: flowering plant in Aug, x 0·08.

Below: leaf, x 0·3. Outer leaflets not symmetrical.

Right: flowers with 2 immature pods, x 1.

Below: immature pod, x 0·75.

39

Runner beans

■□□□□

Scientific name: *Phaseolus coccineus* L. (syn. *P. multiflorus* Willd.).

Decription: Perennial grown as an annual vegetable crop. This species is a vigorous climber with clusters of 20-30 red, red & white, or white flowers on long axillary stalks, followed by long pods. The leaves have 3 leaflets. Field crops have their tops pinched out to check their growth.

Origin & history: Guatemala and neighbouring countries. Introduced to Europe by Spanish explorers. Initially the species was grown as a garden ornamental for its bright red flowers and only in the mid C18th was it used as a vegetable crop in Britain. Runner beans are the most common bean grown in Britain's kitchen gardens and allotments.

Use(s): The unripe seed pods are used as a vegetable.

Sown: May (or for early crops, established from transplants).

Harvested: Late Jul-Oct.

Grown: S England and Wales.

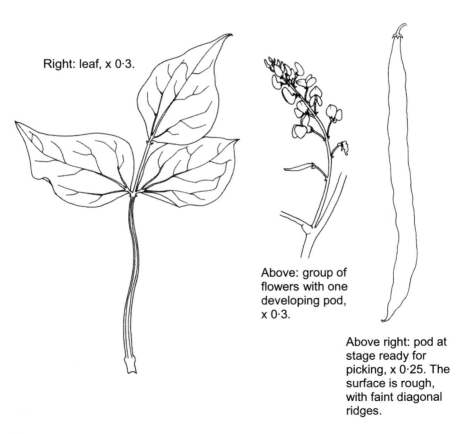

Right: leaf, x 0·3.

Above: group of flowers with one developing pod, x 0·3.

Above right: pod at stage ready for picking, x 0·25. The surface is rough, with faint diagonal ridges.

Soya beans (soybeans)

Scientific name: *Glycine max* (L.) Merr.

Description: Novel (in Britain) annual pulse crop. Up to 1m tall and hairy. The leaves have 3 leaflets. Groups of 3-15 yellowish-white or lilac flowers are followed by conspicuously hairy pods held above the foliage. The leaves die off before harvest, leaving only stems and clusters of pods.

Origin & history: NE China. Possibly selected from the wild species *G. soja* Siebold & Zucc. in the C11th BC. Soya has a very long history of cultivation in Asia. It was known in Europe from the C16th and extensively developed as a field-scale crop in the USA in the C20th. It is now the world's most widely cultivated pulse. Varieties suited to the British climate have only been available to farmers since 1996.

Use(s): The ripe seed of all the British crop is used for human consumption. Many other markets for soya beans are currently served by massive imports, but are being developed for the home-grown crop.

Sown: Late Apr to 1st week May.

Harvested: Early-mid Sep.

Grown: At altitudes of less than 80m above sea level, as far N as York.

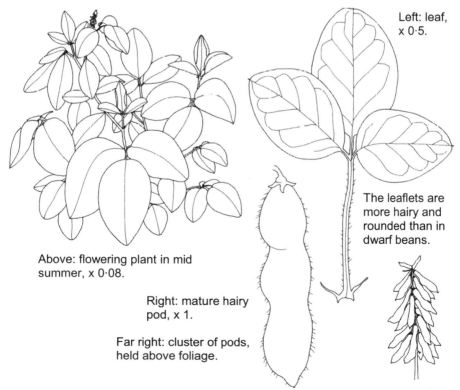

Left: leaf, x 0·5.

The leaflets are more hairy and rounded than in dwarf beans.

Above: flowering plant in mid summer, x 0·08.

Right: mature hairy pod, x 1.

Far right: cluster of pods, held above foliage.

41

Lupins 1

Description: Annual forage or (novel use) pulse crops. All have characteristic 'palmate' leaves with 5-11 identical leaflets, depending on the species. Agricultural lupins are not the same species as ornamental garden lupins.

General origin & history: Mediterranean. Lupins have been in cultivation along the Nile valley and Mediterranean for *c.* 2,000 years. They were formerly used in S England for sheep grazing. These were 'bitter' lupins, containing high levels of alkaloids, which are now used only as green manures. Low alkaloid 'sweet' lupins were bred in the 1930s. White lupins were introduced as a pulse crop in 2001. All the species described below are 'sweet' lupins.

1) White lupin ■□□□□

Scientific name: *Lupinus albus* L.

Use(s): Mature plants are used for whole-crop silage (30% of UK crop) or the dry, protein-rich seeds (40% protein) (remainder of the crop) are used in animal feeds (and can replace soya meal).

Sown: Mid Mar.

Harvested: Early Aug for silage crops. Late Aug-early Sep for seed crops.

Grown: Seed crops S of a line between the Severn and Humber. Forage crops countrywide. This lupin species is the best for chalky soils.

2) Blue (or narrow-leaved) lupin ■□□□□

Scientific name: *L. angustifolius* L. The common name 'narrow-leaved lupin' is more suitable than 'blue lupin', since flower colour can be blue or white. The size and shape of the leaflets are, however, diagnostic.

Use(s): Same as white lupins.

Sown: Early Apr.

Harvested: Same as white lupins.

Grown: Seed crops in N. Forage crops in extreme N of country.

3) Yellow lupin ■□□□□

Scientific name: *L. luteus* L.

Use(s): Ripe seed, as above. Not a forage crop.

Drilled: Early Apr.

Harvested: Early Aug.

Grown: S England on acidic soils unsuitable for other lupin species. Yellow lupins are less commonly planted than the other two lupin species.

Lupins 2

From left: leaves of
white lupin, blue lupin
and yellow lupin,
all x 0·3. White lupin
leaflets have
conspicuous white
hairs along their
margins.

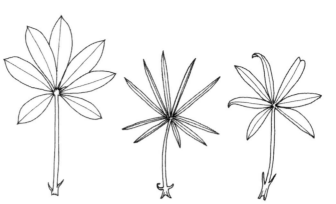

Left: flowering white lupin plant in Jun,
x 0·25. This is the most common lupin
species used agriculturally. The whole
plant is covered by short, downy hairs.

Below, from left: flowers of
white lupin (either white or
pale blue flowers), blue lupin
(either blue or white flowers)
and yellow lupin (flowers
always yellow), all x 0·5.

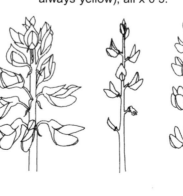

Right: mature white lupin pod, x
0·3. Pods of the other species are
similar but about half the size.
They are held almost upright on
the plant.

Pea family (Leguminosae or Fabaceae)

Clovers and trefoil

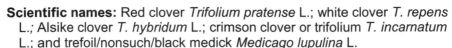

Scientific names: Red clover *Trifolium pratense* L.; white clover *T. repens* L.*;* Alsike clover *T. hybridum* L.; crimson clover or trifolium *T. incarnatum* L.; and trefoil/nonsuch/black medick *Medicago lupulina* L.

Description: Perennials (*T. pratense, repens* and *hybridum*) or annuals (*T. incarnatum* and *M. lupulina*). The leaves have 3 leaflets and the flower heads are comprised of a large number of tightly clustered small flowers.

Origin & history: Red clover is a native species occurring in four forms: wild, broad (introduced from the Netherlands in the C17th), single-cut and late-flowering. White clover is a native species classified into different types according to leaf size. Alsike clover is named after the Swedish village from which it was introduced in the C19th. Crimson clover was introduced from S Europe. Trefoil is a native species.

Use(s): Either used as forage legumes and/or as green manure crops.

Sown: Either undersown into spring cereal crops, or sown in autumn, depending on species.

Harvested: White, red and Alsike clovers are grazed all year. Seed crops are harvested in Sep, sometimes after desiccation. Red clover is now widely used by organic farmers as a green manure crop, which is cut and allowed to rot down *in situ* several times a year.

Grown: Countrywide.

Right: red clover plant x 0·25, leaf x 0·5. It is hairy, has an upright habit, and virtually stalk-less heads of purple flowers. The leaflets usually have a horseshoe mark, and the stipules have purple/red veins.

Not illustrated

White clover: creeping habit; hairless leaflets that usually have horseshoe mark; white flowers.

Alsike clover: hairless leaves; no leaflet mark; the flowers are usually pale pink.

Crimson clover: downy/hairy; no leaflet mark; the flowers are usually crimson, in a pointed head and smothered in long hairs.

Right: trefoil, plant x 0·5, leaf x 1·5. It has a straggling/trailing habit, yellow flowers, black seed pods, and leaflets with a tiny point at tip (unlike species of *Trifolium*).

Lucerne (alfalfa in USA)

■☐☐☐☐

Scientific name: *Medicago sativa* L.

Description: Perennial forage legume. Up to 1m tall, leaves have 3 leaflets, flowers are often blue, pods (where formed) are spiral-shaped.

Origin & history: Caspian. Introduced from Persia to Greece in the C5th BC and grown in Britain since the C17th. The domestication and cultivation of lucerne is associated with the spread and importance of domesticated horses. The plant has a very high nutrient content which makes it ideal for feeding working horses. Lucerne has also been hybridised with the yellow-flowered species, sickle medick (*M. falcata* L.) to increase the crop's hardiness. Lucerne, like sainfoin, has a very deep taproot that gives the plant drought tolerance and allows it to exploit nutrients in the sub-soil beneath the reach of other plants' roots.

Use(s): The foliage is a nutrient-rich feed, used especially for race horses. The crop may be dried indoors (to make barn hay), made into silage or into green-crop dried pellets. The species is not tolerant of grazing.

Sown: Late Apr-mid May. Sometimes undersown into a cereal crop.

Harvested: First cut in mid May, year 2. Subsequent cuts not later than flower-opening stage. Up to 2-3 cuts per season possible in established crops.

Grown: Best suited to SE England. Large crop area in Dengie peninsula of Essex.

Left: part of mature plant, x 0·25.

Right: leaf, x 1. Note the long, toothed stipules.

Sainfoin (cockshead, holy grass) ■☐☐☐☐

Scientific name: *Onobrychis viciifolia* Scop. (syn. *O. sativa* Lam.).

Description: Perennial forage legume. Up to 0·5m tall and covered with sparse downy short hairs. The leaves have 6-12 pairs of leaflets with one smaller terminal leaflet, and red, tube-like stipules. The flowers are in conical heads, and are pink with red veins.

Origin & history: Asia, but now naturalised in Europe. Its history is similar to lucerne. It was introduced by C18th agricultural improvers (e.g. Sir Mordaunt Martin, 'Father of Sainfoin' in Norfolk). The crop was historically important on chalky light land, especially around Newmarket, the Cotswolds and in Hants. Today, it is enjoying a revival on some organic farms and is also included in seed mixtures for roadside verges.

Use(s): The whole plants are used to make nutrient-rich hay or silage, especially for feeding race horses. The aftermath of hay cutting is used for lamb grazing. Sainfoin is also used as a green manure crop and conservation crop.

Sown: Undersown Mar-early May into spring barley crops. 'Common' sainfoin, sometimes grown with a mixture of grasses and white clover, can persist for up to 20 years, but 'giant' sainfoin is short-lived (two years) and is grown alone.

Harvested: Common sainfoin has a single hay cut in May or Jun, and then is used for grazing. Gaint sainfoin has two cuts per annum.

Grown: Mostly in S England.

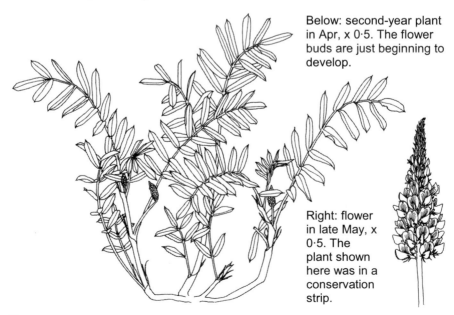

Below: second-year plant in Apr, x 0·5. The flower buds are just beginning to develop.

Right: flower in late May, x 0·5. The plant shown here was in a conservation strip.

Tares (vetches) ■☐☐☐☐

Scientific name: *Vicia sativa* L.

Description: Annual fodder, pulse or green manure crop. The leaves are slightly downy, have 4-8 pairs of leaflets, and end with a branched tendril. Single or paired purple axillary flowers are followed by small, upright, pea-like pods.

Origin & history: A native species, formerly cultivated widely in Britain, with the best crops on lime-rich clay soils. Its importance has declined since the C19th, but is now being revived by organic farmers.

Use(s): The whole plants (in mixture with rye) are used for early sheep grazing, or (in mixture with oats, or field beans and oats) for silage. The ripe, dry seed is used for bird food or fishing bait. The whole plant is also grown as a green manure crop.

Sown: Winter tares Sep-early Oct. Spring tares Feb-Apr.

Harvested: Early spring for sheep grazing. Seed crops harvested end Aug.

Grown: Can be grown in all arable areas.

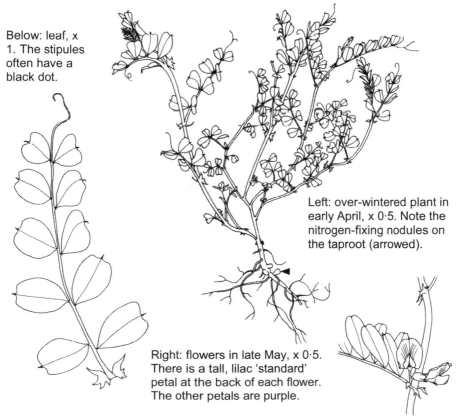

Below: leaf, x 1. The stipules often have a black dot.

Left: over-wintered plant in early April, x 0·5. Note the nitrogen-fixing nodules on the taproot (arrowed).

Right: flowers in late May, x 0·5. There is a tall, lilac 'standard' petal at the back of each flower. The other petals are purple.

Beets 1

Scientific name: *Beta vulgaris* subsp. *vulgaris* (L.) Koch.
Description: Biennials, grown as annual root crops. All have shiny leaves.
General origin & history: Domesticated from the wild sea beet (subsp. *maritima* (L.) Doell) found on UK and continental coasts, initially as a leafy vegetable. Cultivated since at least the time of the Ancient Greeks.

1) Sugar beet

Origin & history: Selected from high-sugar-content fodder beets in C18th Prussia & later developed in France as a European alternative to scarce cane sugar imports during the Napoleonic Wars. It was first commercially processed in 1802 in Prussia, but only uninterruptedly cultivated in Britain since 1920. The crop now provides half of our sugar requirements.
Use(s): The roots are processed to extract sugar (residues are used for animal feed). The leaves may be saved for cattle feed. Roots also have potential for bio-ethanol production (liquid fossil fuel substitute/additive).
Sown: 20th Mar-10th Apr.
Harvested: Sep-Nov. Lifted roots are piled for protection from winter frosts.
Grown: On contract to British Sugar plc., near processing factories at Allscott (Shropshire), Bury St Edmunds (Suffolk), Cantley (Norfolk), Newark, Wissington (Norfolk) and York.

2) Mangels (mangolds) & fodder beet

Origin & history: First grown as a field crop in C17th Spain. Mangels (mangel wurzels: yellow-fleshed roots with low dry-matter content) have been grown in Britain since the late C18th. Fodder beets (high dry-matter hybrids of mangolds and sugar beet) were introduced after 1940.
Use(s): Roots as animal feed. Roots and leaves of fodder beet for feed.
Sown: Late Apr.
Harvested: Oct-Nov.
Grown: Especially in dairying districts in S & SW.

3) Beetroot (table beet or red beet)

Origin & history: Beets with sweet-tasting red roots were first described in 1538. They were introduced to Britain from Italy during the late C16th.
Use(s): A root vegetable for immediate use and for pickling.
Sown: Mar onwards. Early crops are sown under plastic film.
Harvested: 'Bunching' crop Jun-Jul (small bunches of young plants for salad use). Maincrop Jul-mid Nov for processing.
Grown: Large area in the Fens.

Beets 2

Below: mature sugar beet plant, x 0·1. A mass of rootlets emerge from the so-called 'sugar groove' running down the side of the root. The roots are always conical in shape. Soil level (on plant diagrams on this page) is shown by the arrow.

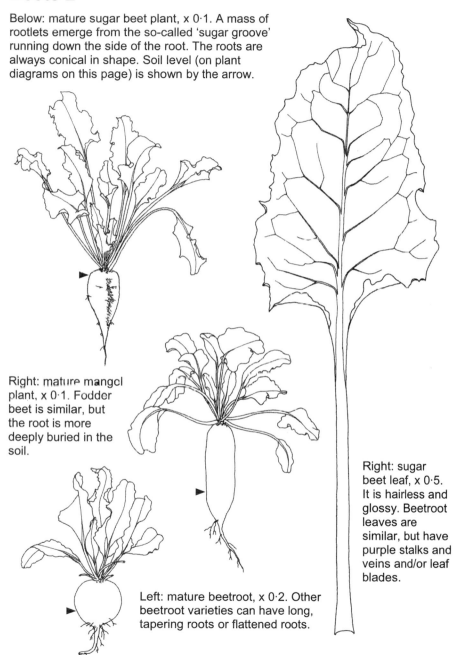

Right: mature mangel plant, x 0·1. Fodder beet is similar, but the root is more deeply buried in the soil.

Right: sugar beet leaf, x 0·5. It is hairless and glossy. Beetroot leaves are similar, but have purple stalks and veins and/or leaf blades.

Left: mature beetroot, x 0·2. Other beetroot varieties can have long, tapering roots or flattened roots.

Potatoes

■■■■□

Scientific name: *Solanum tuberosum* L.

Description: Perennial, grown on ridges as an annual vegetable crop. It has white or mauve flowers with 5 fused petals, and divided hairy leaves.

Origin & history: S & C America. Many different *Solanum* species were in cultivation as early as the 3rd millenium BC. The 'Peruvian' potato (*S. andigena* Juz. et Buk.) was the first species introduced to Europe (Spain 1570, Britain 1590), but was only widely accepted as a vegetable 200 years later. *S. tuberosum* was introduced from Chile in the 1840s.

Above: foliage ('haulm') in early May, x 0·1.

Use(s): The tubers are used for human consumption ('ware' crops). These are divided into earlies (new, or salad, potatoes), 2nd earlies, and maincrop potatoes (the majority of the crop: used for crisps and frozen chips and ordinary & baking potatoes). 'Seed' crops are also grown to multiply stocks of each variety for future planting.

Planted: Earlies Jan & Feb. 2nd earlies Mar. Maincrops late Mar in S, Apr in N.

Harvested: Earlies 2nd wk Jun onwards. 2nd earlies Jul & Aug. Maincrop late Sep onwards. All the crop is lifted before winter and put into cold storage.

Grown: Earlies especially in W Cornwall, S Pembrokeshire, E Kent and Jersey. Seed crops in Scotland and N Ireland. Others countrywide on suitable soils.

Right: leaf x 0·25.

Right: flowers x 0·5. White or mauve fused petals and conspicuous yellow stamens.

Linseed (including linola) and flax

Scientific name: *Linum usitatissimum* L.

Description: Annual oilseed (linseed) or fibre (flax) crop. Up to 0·7m tall, slender stemmed, with sky-blue (or in some varieties, white) flowers.

Origin & history: SW Asia. In cultivation for *c.* 5,000 years. In S areas, the crop developed into linseed types (for oilseed production) and in N areas, it developed into flax (for long stems as a fibre crop). Flax was formerly widely cultivated, especially in Ireland, for linen production.

Use(s): The ripe seed of conventional linseed (with dark brown seeds) is crushed for oil, rich in linolenic acid; used for paints, varnishes and Linoleum manufacture. Seed oil of linola (so-called 'edible linseed'; with golden seeds) is rich in linoleic acid and is used as a culinary oil. Both types of unprocessed seed are used as a health food and in breads. The seed meal is a valuable animal feed. The mature stems of flax yield an industrial fibre, though very little is currently grown in Britain due to a lack of processing facilities.

Sown: Mid Mar-mid Apr.

Harvested: Late Aug-Sep.

Grown: Countrywide.

Left: flowering plant in June, x 0·2.

Left: underside of leaf showing three veins, x 2.

Left: newly opened flower, x 1.

Left: mature seed capsule, side view, x 1.

Carrot

■■□□□

Scientific name: *Daucus carota* subsp. *sativus* (L.) Scheubler & Martens.

Description: Biennial, grown as an annual root crop. It has feathery, fragrant foliage and an orange taproot (garden novelty varieties have purple, yellow, or white roots). Root shape varies according to variety.

Origin & history: Afghanistan. Original domesticated carrots (*c.* 2,000 years ago) were purple-rooted. Selections were then made for orange and yellow types. Orange types were fully developed in the Netherlands in the C17th-18th, then introduced to Britain. White cattle-carrots were formerly grown for stock feed. Wild carrots (subsp. *carota*) also occur in Britain.

Use(s): The root is a vegetable for human consumption. Reject roots are used for cattle feed.

Sown: Jan-Jul for a succession of crops. Early crops sometimes sown under polythene, which can give a fortnight's advantage in growing time. Over-wintering crops sown in Oct.

Harvested: Approximately 70-100 days after emergence of seedlings. Roots of late crops protected from frosts by a thick covering of straw in the field.

Grown: On deep, sandy, well-drained soils.

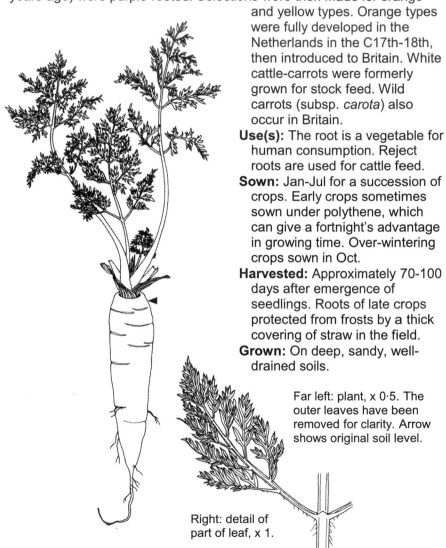

Far left: plant, x 0·5. The outer leaves have been removed for clarity. Arrow shows original soil level.

Right: detail of part of leaf, x 1.

Parsnip ■□□□□

Scientific name: *Pastinaca sativa* L.

Description: Biennial, grown as an annual root crop. The plants have mid-green, divided hairless leaves, and a conical white taproot.

Origin & history: The wild species exists in Britain as a grassland plant of chalky soils. It has been cultivated since Roman times, but improved forms have only existed since the Middle Ages. Formerly it was also grown for cattle feed on soils too heavy for cattle-carrots. Parsnips are dwindling in popularity in continental Europe, but not in Britain.

Use(s): The root is a vegetable for human consumption, available all year round, but with the strongest demand in winter.

Sown: Jan-May.

Harvested: Oct to following Mar.

Grown: On deep, sandy, well-drained soils.

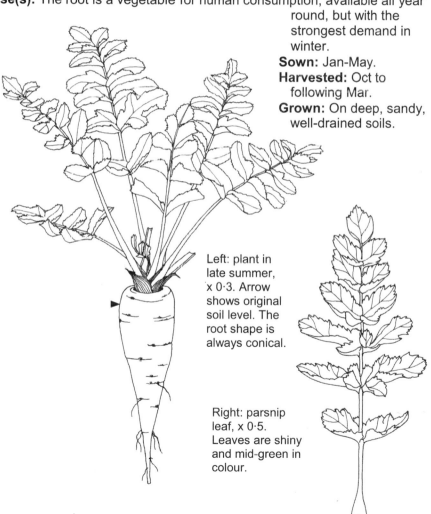

Left: plant in late summer, x 0·3. Arrow shows original soil level. The root shape is always conical.

Right: parsnip leaf, x 0·5. Leaves are shiny and mid-green in colour.

Celery

■□□□□

Scientific name: *Apium graveolens* L. var. *dulce* (Miller) Pers.

Description: Biennial, grown as an annual vegetable. There are two types: self-blanching (mainly green stems) and trench celery (white or pinkish stems); the latter is more winter-hardy. Celery has hairless, shiny, divided leaves on stout fleshy stalks.

Origin & history: Wild celery (smallage) is a waterside species in Britain and Europe. Several celery-flavoured or celery-like plants were used as potherbs in the Middle Ages and earlier. Cultivated celery was developed in C17th Italy from smallage, which tastes bitter. The var. *rapaceum* (Miller) Gaudich. is celeriac, a form grown for its knobbly, enlarged root.

Use(s): The fleshy leaf stalks of celery are used as a salad vegetable.

Sown: Mar & Apr for transplanting May & Jun.

Harvested: Sep to following Feb.

Grown: Mainly in the Fens.

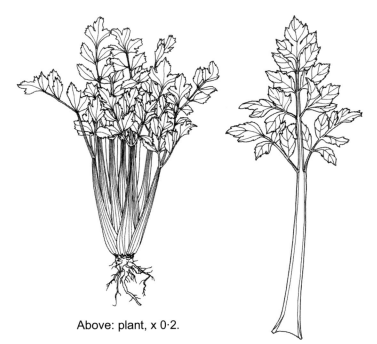

Above: plant, x 0·2.

Above: leaf, x 0·3.

Lettuce

Scientific name: *Lactuca sativa* L.

Description: Annual salad crop. In its non-flowering stage, it is low-growing, with variable leaf shapes and colours depending on variety.

Origin & history: S Mediterranean. Cultivated in Egypt in the 5th millenium BC, possibly as an oilseed crop. Cos and cabbage types were grown by the Ancient Persians, and 11 different types by the Romans. Modern types include cabbage lettuce (butterhead and crisphead: var. *capitata* L.), cos lettuce (var. *romana* Gars.), loose-leaf lettuce (var. *crispa* L.) and oriental 'celtuce' (var. *angustana* Irish), which is grown in gardens for its succulent elongated stem.

Use(s): Leaves of loose-leaf types, and whole heads of hearting types are used as a salad vegetable.

Sown: In modules in glasshouses and transplanted 3 wks later.

Harvested: Loose-leaf lettuce around 50 days after transplanting; head lettuce around 75 days post-transplanting.

Grown: Widely in the Fens.

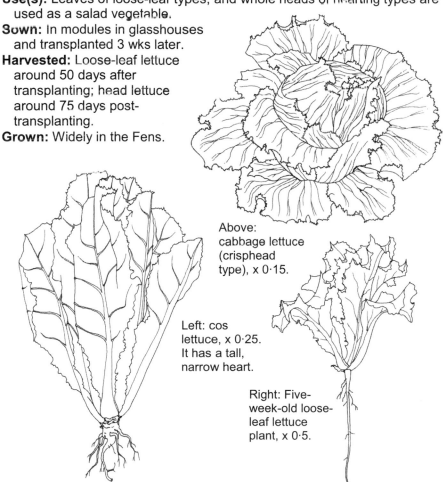

Above: cabbage lettuce (crisphead type), x 0·15.

Left: cos lettuce, x 0·25. It has a tall, narrow heart.

Right: Five-week-old loose-leaf lettuce plant, x 0·5.

Sunflower

■□□□□

Scientific name: *Helianthus annuus* L.

Description: Annual seed or oilseed crop. Up to 1·5-2m tall (garden forms taller), with massive flower head of central brown or yellow disc florets surrounded by outer yellow ray florets. The plant has large hairy leaves.

Origin & history: N America. Domesticated by native Americans from wild, small-seeded forms and introduced to the Old World in 1616 to become a familiar garden ornamental. Sunflowers were first cultivated as a field-scale oilseed crop in Russia or Germany in the C18th. Only in recent years have new varieties been developed that reliably crop in S England. Sunflowers have also been evaluated for silage-making.

Use(s): The ripe seed is mainly used as bird seed. The seed could be crushed for home-grown sunflower oil, but there is no crushing capacity and the crop area is too small at present. Experimental pressing of British seed has yielded an elite culinary oil. Whole plants are also used in game cover mixtures.

Sown: Apr & early May.

Harvested: Sep & Oct.

Grown: SE of a line between the Wash and E Dorset.

Left: flowering plant, x 0·1.

Right: leaf (hairy/bristly), x 0·5.

Calendula (pot marigold)

Scientific name: *Calendula officinalis* L.

Description: Perennial or annual novel industrial oilseed crop. Up to 0·75m tall, with mid-green foliage, hairy angular stems and yellow/orange daisy-like flowers, followed by tightly packed heads of C-shaped spiny seeds.

Origin & history: S Europe and E Mediterranean. Long in cultivation as a garden herb with many traditional culinary and medicinal uses. Calendula has recently been developed as a novel industrial oilseed crop.

Use(s): The ripe seed is crushed for oil, rich in calendic acid; the most rapidly-drying fatty acid known in nature. This has potential uses in the paints and coatings industry, in some industrial nylon products and as a home-grown alternative to imported tung oil, the active component in fast-drying products like yacht varnish and inks.

Sown: Late Mar-late Apr (potential for winter crops in S England).

Harvested: Late Aug.

Grown: Commercialised in 2006.

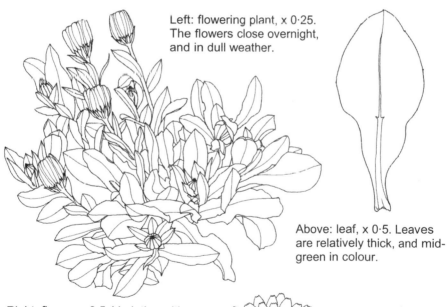

Left: flowering plant, x 0·25. The flowers close overnight, and in dull weather.

Above: leaf, x 0·5. Leaves are relatively thick, and mid-green in colour.

Right: flower, x 0·5. Varieties with double flowers are also grown. The flowers have a strong characteristic smell.

Far right: unripe seed head in late May, x 0·5. At this stage the seeds are still green and tightly folded in.

Industrial hemp

■ □ □ □ □

Scientific name: *Cannabis sativa* L.

Description: Annual industrial fibre crop. Up to 3m tall, with divided leaves that have 3-11 toothed leaflets. The flowers are green, with male and female flowers on separate plants.

Origin & history: Long history of cultivation. Two races were developed from the wild species; in S latitudes for narcotic production (subsp. *indica*) and in N latitudes for fibre production (subsp. *sativa*). The latter was once extensively grown in the Fens for processing into sailing ships' cables/ropes etc. Production in Britain stopped in the 1950s because of new drug legislation. The crop was recently revived after industrial hemp varieties which only contain minute amounts of the narcotic compound (too little for drug use) were bred.

Use(s): The stems yield (i) fibre; processed into cigarette and archive-quality papers, interior car door panels, and building insulation and (ii) an inner pithy core ('shiv'), used for horse bedding. The ripe seed (from a short-strawed variety) is used for bird food, fishing bait, or is crushed for oil for human consumption.

Sown: Around 3rd wk Apr.

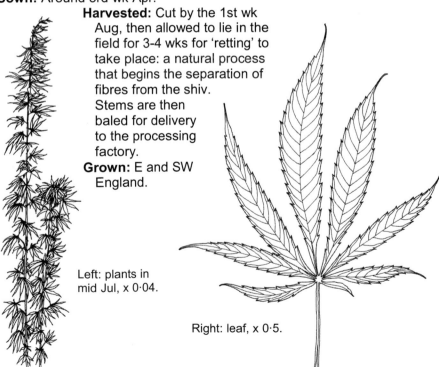

Harvested: Cut by the 1st wk Aug, then allowed to lie in the field for 3-4 wks for 'retting' to take place: a natural process that begins the separation of fibres from the shiv. Stems are then baled for delivery to the processing factory.

Grown: E and SW England.

Left: plants in mid Jul, x 0·04.

Right: leaf, x 0·5.

Borage

■☐☐☐☐

Scientific name: *Borago officinalis* L.

Description: Annual pharmaceutical oilseed crop. Up to 0·5m tall, coarse, bristly/hairy, with royal blue or violet flowers.

Origin & history: Mediterranean. Long in cultivation, initially in Turkish Asia and Syria, then in Moorish Spain, as a garden herb. Introduced to Britain by the Romans but first grown as a field crop only in 1983.

Use(s): The ripe seed is crushed for oil, rich in gamma-linolenic acid (GLA), used for dietary supplements (sold as starflower oil). Around half the world's supply of borage oil comes from Yorkshire farms. Borage contains twice as much GLA as evening primrose (*Oenothera biennis* L.), the other main oilseed GLA source, grown in China and no longer in Britain.

Sown: Early-mid Apr.

Harvested: End Aug.

Grown: Mainly in Suffolk, Essex, Yorks. & N Lincs.

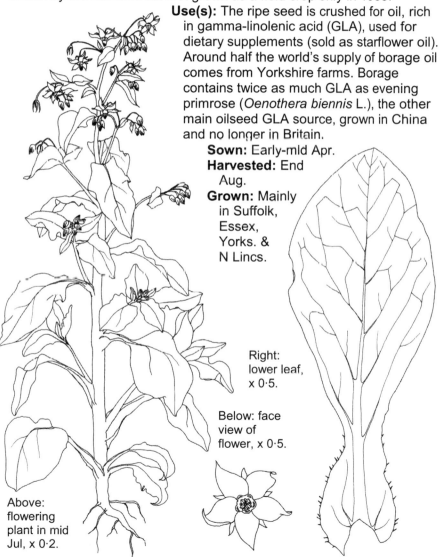

Right: lower leaf, x 0·5.

Below: face view of flower, x 0·5.

Above: flowering plant in mid Jul, x 0·2.

Echium

■☐☐☐☐

Scientific name: *Echium plantagineum* L.

Description: Annual pharmaceutical oilseed crop. Up to 1m tall, coarse, multi-stemmed and sprawling. Individual plants may have white, blue or purple flowers, so there is a mixture of colours in echium fields.

Origin & history: Europe including Britain. *Echium* species are wildflowers on sandy soils and have many traditional herbal uses. Selected varieties are used as garden ornamentals. Echium was recently developed as a novel oilseed and has been grown in Britain as an arable crop for only about 5 years. Its classification is unclear; considered by some to be a synonymn of *E. lycopsis* L. and/or *E. vulgare* L.

Use(s): The ripe seed is crushed for oil, rich in stearidonic acid, used in skin and sun creams for its anti-wrinkle properties.

Sown: Early-mid Apr.

Harvested: Late Aug.

Grown: As far N as Yorkshire.

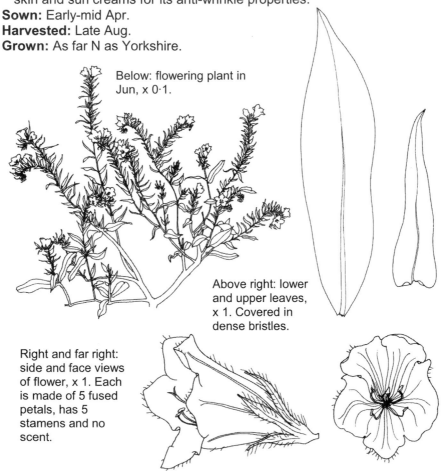

Below: flowering plant in Jun, x 0·1.

Above right: lower and upper leaves, x 1. Covered in dense bristles.

Right and far right: side and face views of flower, x 1. Each is made of 5 fused petals, has 5 stamens and no scent.

Culinary (or oil) poppy

■☐☐☐☐

Scientific name: *Papaver somniferum* L.

Description: Annual seed crop. Up to 1m tall with grey, waxy leaves and large, showy, mauve, short-lived flowers with 4 petals and many black stamens, followed by large seed capsules containing tiny black seeds.

Origin & history: SW Asia. Domesticated in the Neolithic period, grown by the Ancient Greeks and spread to India and China by the C18th. It is unparalleled as a drug plant, being the source of opium (20 different alkaloids including morphine and codeine). A vast illegal crop is grown in Afghanistan, destined for heroin production. The drug content of different varieties varies. Low-opium types are grown for seed (the seeds are drug-free). Some varieties are used as garden ornamentals.

Use(s): The ripe seed (maw seed) is used to decorate and flavour speciality bakery products. The seed can also be crushed for oil. The first pressing gives a clear edible oil (olivette) and the second pressing with heat gives a coloured oil used in artists' paints.

Sown: Spring.

Harvested: Late summer.

Grown: Small crop in Wilts. and Hants.

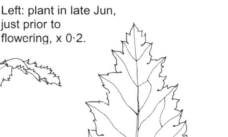

Left: plant in late Jun, just prior to flowering, x 0·2.

Above: mature seed capsule, x 0·5.

Left: lower leaf, x 0·5, typically grey and waxy, sometimes with sparse hairs on the midrib.

Game cover and conservation crops 1

Straight stands or mixtures of various species are sown in strips along field headlands, or in blocks near woods, to provide cover and winter food (usually seeds) for game birds and wildlife.

Grass species used

Maize, millet, miscanthus, triticale and:

1) Canary grass or canary seed (*Phalaris canariensis* L.). An introduced annual with club-like seed heads. This species is also occasionally grown on a very small scale for bird seed.
2) Reed canary grass (*Phalaris arundinacea* L.). A tall, native perennial of lowland wet areas. This species is also under evaluation for biomass production, like miscanthus.
3) Sorghum (*Sorghum bicolor* (L.) Moench Caudatum group), including milo (Subglabrescens group) and giant sorghums, which are hybrids with Sudan grass (*Sorghum sudanense* (Piper) Stapf.). Annual cereals originating from Africa. Game cover crops can produce ripe seed in good seasons.

Broadleaved species used

Fodder beet, fodder radish, fodder rape, gold-of-pleasure (camelina), kale, linseed, lucerne, sainfoin, stubble turnips, sunflower, tic beans, white mustard, and:

1) Buckwheat (*Fagopyrum esculentum* Moench). A straggling annual with small pink flowers followed by pyramid-shaped seed.
2) Jerusalem artichoke (*Helianthus tuberosus* L.). A perennial, surviving the winter as tubers. Tall, with yellow, sunflower-like flowers.
3) Phacelia (*Phacelia tanacetifolia* Benth.). A hairy annual with fern-like leaves and small, sweetly-scented, lilac, 5-petalled flowers, borne in a double row on a curling flower head. It is a good bee plant.
4) Quinoa (*Chenopodium quinoa* Willd.). An annual, somewhat similar to the wild species fat-hen (*C. album* L.). It produces a vast amount of tiny nutritious seed in heads tinged red, orange or yellow.
5) 'Stand and deliver' (*Cichorium* sp.). A perennial forming a rosette of dandelion-like leaves in its first year. It bolts from the 2nd year onwards to 2m tall, with blue flowers.
6) Texsel/Texel greens (*Brassica carinata* A. Braun). A fast-growing annual brassica with small, glossy leaves. It is only grown as far N as York.
7) Yellow sweet clover (*Melilotus officinalis* (L.) Pallas). A tall, slender, branching biennial, its leaves have 3 leaflets, and the flowers are yellow.

Game cover and conservation crops 2

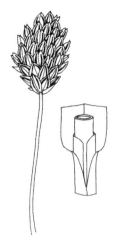

Above: canary grass ear (x 0·5) and leaf base (x 3).

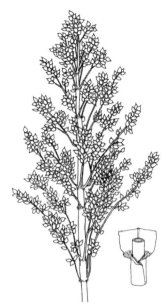

Above: sorghum ear and leaf base, both x 0·5.

Above: quinoa, top of stem bearing ripe seed, x 0·1. The seed clusters and leaves turn shades of orange and red in autumn.

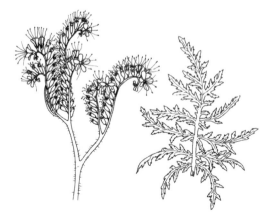

Above: phacelia flower head and leaf, x 0·5.

Right: yellow sweet clover, tip of flowering stem, x 0·25.

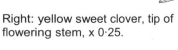

Glossary

Aftermath Short stems of living crop plants that remain after hay or silage has been harvested.

Alkaloid General term for a plant-derived compound with a bitter taste that has powerful effects (medicinal or poisonous) on animals and humans.

Amino acids Group of compounds that are the building blocks of proteins. Essential amino acids must be derived from the diet, but non-essential amino acids can be made inside the body.

Annual Plant that germinates, flowers, sets seed, then dies in a single growing season.

Awn Bristle or hair-like structure.

Axil Angle between a leaf and its stem.

Axillary bud Bud which is present in a leaf axil. If this buds develops, it will grow into a new branch or a cluster of flowers.

Biennial Plant that completes its life cycle in two years, flowering and setting seed in the second year only. It often survives the winter by building up food reserves in a fleshy root, e.g. carrots, parsnips etc.

Biomass Plant- or animal-derived material. Important as a renewable fuel in power stations, e.g. miscanthus canes.

Brassica Plant in the genus *Brassica*, which includes cabbages, cauliflowers, Brussels sprouts etc.

Cereal Plant in the grass family whose seeds are used as food, e.g. wheat.

Compound leaf Leaf comprising a number of leaflets, e.g. in clover.

Cotyledon or seedling leaf. The first leaf or leaves produced by a germinating seedling. Seedlings may produce one cotyledon (Monocotyledons: grasses, lilies etc), or two cotyledons (Dicotyledons: broadleaved plants). Cotyledon(s) are often different in appearance from subsequent true leaves.

Crucifer Plant in the family Cruciferae (also called Brassicaceae). So called because the flowers have 4 petals in the shape of a cross.

Calendic acid Fatty acid found in the seed oil of *Calendula officinalis*. It has potential industrial uses in paints etc.

Desiccate To hasten seed-ripening in a crop by prematurely killing the foliage with a herbicide.

Ear General term for the flowering head of a grass or cereal.

Erucic acid Fatty acid found in the seed oil of various crucifer species, which has various uses in the plastics industry.

Fatty acids Group of compounds that are the building blocks of fats and oils. Linolenic acids and linoleic acid are known as essential fatty acids.

They must be derived from the diet and cannot be made inside the body.

Floret Small flower, present in e.g. grasses and plants in the daisy family. In the latter, the 'flower' consists of many tubular disc florets in the centre, surrounded by an outer ring, or rings, of strap-like ray florets, commonly called the 'petals'.

Fodder Plant material harvested for feeding to animals, e.g. fodder beets.

Forage Crop grown to be grazed by animals *in situ*. Note that the terms fodder and forage are often used interchangeably.

Game cover Mixtures, or single species, planted in strips or blocks, especially near small woods to provide food and shelter for game birds. They also benefit wild species.

Glucosinolates Group of nitrogen and sulphur-containing compounds found in crucifers (cabbages etc.). They are important for several reasons: some glucosinolates are responsible for taste (e.g. of Brussels sprouts), some have potential anti-cancer properties (e.g. in broccoli) and some are so-called 'anti-nutritional factors' that reduce the suitability of certain products for animal feed (e.g. oilseed rape meal; hence the development of varieties low in glucosinolates).

Grain In the narrow sense, the seeds of cereals. In the wider sense, also used to describe other small seeds used as food, e.g. buckwheat.

Green manure Short-lived crop, e.g. mustard, sown to maintain and/or increase soil fertility. It is ploughed into the soil, before maturity or flowering, where it rots down to increase the levels of soil organic matter.

Herbage Pasture grasses eaten by animals.

Hulled Cereal grain tightly enclosed by a tough, adherent husk.

Industrial crop Crop not intended to enter the food chain, e.g. fibre crops (hemp & flax) or certain oilseed crops whose seed oil is inedible but has valuable industrial applications.

Leaflet Constituent part of a compound leaf, e.g. a clover leaf comprises three leaflets. Often found in plants in the family Leguminosae (Fabaceae). Individual leaflets do not have axillary buds associated with them, unlike true leaves.

Legume Plant in the family Leguminosae (Fabaceae). Often has compound leaves, characteristic sweetpea-like flowers, and root nodules containing bacteria that can convert atmospheric nitrogen into a form the plant can use. As nitrogen is one of the chief components of proteins, legumes are highly valuable sources of dietary protein.

Linoleic acid Essential fatty acid, first isolated in 1844.

Linolenic acid Essential fatty acid that occurs in two forms: alpha-linolenic acid and gamma-linolenic acid (GLA). Alpha-linolenic acid is common in linseed oil, and used as a drying oil e.g. in paints. GLA helps regulate hormonal systems and lowers blood pressure.

Naked Cereal grain not tightly enclosed by an adherent husk, making it more palatable for human and animal feedstuffs.

Nervonic acid Fatty acid important in brain and nervous system function.

Oilseed Crop grown for its oil-rich seeds.

Panicle Type of flower head typified by oat and millet ears.

Perennial Plant that continues its growth from year to year.

Pulse Edible seeds of plants from the Leguminosae family, e.g. peas.

Rhizome Horizontal, creeping, underground stem.

Silage Preserved foliage of grass or other crop (whole-crop silage) used for animal feed. Made in large covered heaps or plastic-wrapped bales by excluding air and encouraging fermentation by lactic acid bacteria.

Spike Type of grass or cereal ear comprising stalk-less spikelets or florets, e.g. wheat.

Spikelet Sub-unit of spike, comprising florets, e.g. in rye each spikelet has two fertile florets that each develop a grain.

Spring crop Crop sown in spring (see winter crops).

Stamen Part of a flower that produces pollen.

Stearidonic acid Fatty acid widely found in fish oil and also in echium seed. It has anti-inflammatory effects.

Stipule Outgrowth at the base of a leaf, e.g. in peas.

Stubble Severed stems of dead annual crop left standing after harvesting.

Style Elongated outgrowth of the female part(s) of a flower that ends in the stigma, the receptive surface which receives pollen during pollination.

Subsp. (subspecies). Main division of a species.

Swathing To cut a crop and leave it in the field to dry, lying on its own stubble. Often used to hasten seed ripening.

Syn. (synonym). Alternative name of a plant, not accepted as the valid one.

Tendril Slender outgrowth which helps support the plant by twining around neighbouring stems, e.g. in peas.

Terminal At the end of e.g. a terminal leaflet on a leaf.

Threshing Process by which ripe seed is separated from the rest of the plant and partially cleaned of debris. Carried out by a combine harvester (combine).

True leaves Leaves that develop after the cotyledons.

Undersown Where the seed of one crop is sown into another crop, which has been established earlier in the year and which is still growing.

Var. (varietas or botanical variety). A division in plant classification that is smaller than a subspecies.

Winter crop Crop sown in autumn, which over-winters to be harvested the following year. Usually higher yielding than its spring counterpart.

Index of common names

Page numbers in bold indicate illustrations.

Index of scientific names

Page numbers in bold indicate illustrations. Names in brackets are synonyms.

Notes

Notes